Chiropractic

Edzard Ernst

Chiropractic

Not All That It's Cracked Up to Be

 Springer

Edzard Ernst
Cambridge, UK

ISBN 978-3-030-53117-1   ISBN 978-3-030-53118-8   (eBook)
https://doi.org/10.1007/978-3-030-53118-8

This Springer imprint is published by the registered company Springer Nature Switzerland AG
The registered company address is: Gewerbestrasse 11, 6330 Cham, Switzerland

Contents

Founder of Chiropractic. The Creator of Chiropractic Science. The Originator of
Vertebral Adjusting. The Developer of Chiropractic Philosophy. The Foun-
tain Head of the Principles of Chiropractic, their skillful application
for the use of humanity and the reasons why and how they
Govern Life in Health and Disease. Lecturer and
Demonstrator on the Science, Art and
Philosophy of Chiropractic.

Fig.-Daniel David Palmer; source: US National Library of Medicine

1

Introduction

"*There is no alternative medicine. There is only scientifically proven, evidence-based medicine supported by solid data or unproven medicine, for which scientific evidence is lacking.*" These words of Fontanarosa and Lundberg were published 22 years ago.[1] Today, they are as relevant as ever, particularly to the type of healthcare I often refer to as 'so-called alternative medicine' (SCAM),[2] and they certainly are relevant to chiropractic.

Invented more than 120 years ago by the magnetic healer DD Palmer, chiropractic has had a colourful history. It has now grown into one of the most popular of all SCAMs. Its general acceptance might give the impression that chiropractic, *the art of adjusting by hand all subluxations of the three hundred articulations of the human skeletal frame,*[3] is solidly based on evidence. It is therefore easy to forget that a plethora of fundamental questions about chiropractic remain unanswered.

I wrote this book because I feel that the amount of misinformation on chiropractic is scandalous and demands a critical evaluation of the evidence. The book deals with many questions that consumers often ask:

- How well-established is chiropractic?
- What treatments do chiropractors use?
- What conditions do they treat?

---

[1] Fontanarosa PB, Lundberg GD. Alternative medicine meets science. *JAMA*. 1998;280(18):1618–1619. http://doi.org/10.1001/jama.280.18.1618.

[2] Ernst E. *SCAM: So-Called Alternative Medicine.* Societas (2018).

[3] Palmer DD. *Text-Book of the Science, Art and Philosophy of Chiropractic*. Rev. edn. Echo Point Books & Media (2019).

- What claims do they make?
- Are their assumptions reasonable?
- Are chiropractic spinal manipulations effective?
- Are these manipulations safe?
- Do chiropractors behave professionally and ethically?

Am I up to this task, and can you trust my assessments? These are justified questions; let me try to answer them by giving you a brief summary of my professional background.

I grew up in Germany where SCAM is hugely popular. I studied medicine and, as a young doctor, was enthusiastic about SCAM. After several years in basic research, I returned to clinical medicine, became professor of rehabilitation medicine first in Hanover, Germany, and then in Vienna, Austria. In 1993, I was appointed as Chair in Complementary Medicine at the University of Exeter. In this capacity, I built up a multidisciplinary team of scientists conducting research into all sorts of SCAM with one focus on chiropractic. I retired in 2012 and am now an emeritus professor. I have published many peer-reviewed articles on the subject, and I have no conflicts of interest. If my long career has taught me anything, it is this: in the best interest of consumers and patients, we must insist on sound evidence; not opinion, not wishful thinking; evidence.

In critically assessing the issues related to chiropractic, I am guided by the most reliable and up-to-date scientific evidence. The conclusions I reach often suggest that chiropractic is not what it is often cracked up to be. Hundreds of books have been published that disagree. If you are in doubt who to trust, the promoter or the critic of chiropractic, I suggest you ask yourself a simple question: who is more likely to provide impartial information, the chiropractor who makes a living by his trade, or the academic who has researched the subject for the last 30 years?

This book offers an easy to understand, concise and dependable evaluation of chiropractic. It enables you to make up your own mind. I want you to take therapeutic decisions that are reasonable and based on solid evidence. My book should empower you to do just that.

May 2020, Edzard Ernst.

2

The History of Chiropractic

I was the first to adjust the cause of disease.
— D. D. Palmer

Who better to summarise the beginnings of chiropractic than Daniel David Palmer, the man who never left any doubt that it was he, and he alone, who invented it? Here is the start of his article[1] entitled 'A Brief History of the Author and Chiropractic':

> I was born on March 7, 1845, a few miles east of Toronto, Canada. My ancestors were Scottish and Irish on my maternal and English and German on my paternal side.
>
> When my grandparents settled near the now beautiful city of Toronto, there was but one log house, the beginning of that great city. That region was then known as 'away out west'.
>
> I came within one of never having a mamma. My mother was one of a pair of twins one of which died. The one who lived weighed only one and a half pounds.
>
> When a baby I was cradled in a piece of hemlock bark. My mother was as full of superstition as an eff is full of meat, but my father was disposed to reason on the subjects pertaining to life.

---

[1] Palmer DD. *Text-Book of the Science, Art and Philosophy of Chiropractic*. Rev. edn. Echo Point Books & Media (2019).

© Springer Nature Switzerland AG 2020
E. Ernst, *Chiropractic*,
https://doi.org/10.1007/978-3-030-53118-8_2

I was a magnetic healer for nine years previous to discovering the principles which comprise the method known as Chiropractic. During this period much of what was necessary to complete the science was worked out. I had discovered that many diseases were associated with derangements of the stomach, kidneys and other organs…

One question was always uppermost in my mind in my search for the cause of disease. I desired to know why one person was ailing and his associate, eating at the same table, working in the same shop, at the same bench, was not. **Why?** What difference was there in the two persons that caused one to have pneumonia, catarrh, typhoid or rheumatism, while his partner, similarly situated, escaped? Why? This question had worried thousands for centuries and was answered in September 1895…

September 18, 1895 is the day when Daniel David (DD) Palmer invented chiropractic. On that occasion, he manipulated the spine of a deaf janitor by the name of Harvey Lillard, allegedly curing him of his deafness (Box 2.1). *An examination showed a vertebra racked from its normal position,* Palmer wrote*, I reasoned that if that vertebra was replaced, the man's hearing should be restored* (see footnote 1). Palmer's second patient was a man suffering from heart disease (Box 2.2).[2] Palmer wrote: *I examined the spine and found a displaced vertebra pressing against the nerves which innervate the heart. I adjusted the vertebra and gave immediate relief…* (see footnote 1)

Palmer had been one of 6 children of parents who had immigrated from Canada to the US in search for work. During the first 20 years of his adult life, he worked in various professions (Box 2.3). His attraction to all things medical then made him try his luck as a magnetic healer,[3] and it was then that the fateful encounter with his janitor changed his life for ever.

Palmer later wrote in his book, 'Chiropractor's Adjuster',[4] that he learned about manipulation during a séance from the ghost of a medical practitioner named Jim Atkinson whose work, 50 years earlier, had formed the basis of Palmer's new method:

The knowledge and philosophy given to me by Dr. Jim Atkinson, an intelligent spiritual being, together with explanations of phenomena, principles resolved from causes, effects, powers, laws, and utility, appealed to my reason. The

---

[2] Senzon SA. The Chiropractic Vertebral Subluxation Part 2: The Earliest Subluxation Theories From 1902 to 1907. *J Chiropr Humanit*. 2019;25:22–35. Published 2019 Apr 6. http://doi.org/10.1016/j.echu.2018.10.009.

[3] Ernst E. Chiropractic: a critical evaluation. *J Pain Symptom Manage*. 2008;35(5):544–562. http://doi.org/10.1016/j.jpainsymman.2007.07.004.

[4] Palmer DD. *The chiropractic adjuster: A compilation of the writings of D.D. Palmer*. Palmer School of Chiropractic (1921).

method by which I obtained an explanation of certain physical phenomena, from an intelligence in the spiritual world, is known in biblical language as inspiration. In a great measure, The Chiropractor's Adjuster was written under such spiritual promptings.

Palmer stated that *chiropractic was not evolved from medicine or any other method, except that of magnetic* (see footnote 2). A friend of Palmer's, Rev Samuel Weed, is credited with creating the word 'chiropractic', but Palmer coined the term "innate intelligence" (or "innate") for the assumed "energy" or "vital force," which, according to his belief, enables the body to heal itself. Palmer claimed that the "innate" controls all body functions. In the presence of "vertebral subluxation," the innate was blocked, he postulated. Thus, subluxations are the cause of all disease. Palmer developed spinal manipulations to correct subluxations and thus unblock the flow of the innate and defined chiropractic as *a system of healing based on the premise that the body requires unobstructed flow through the* nervous *system of innate intelligence* (see footnote 2).

Palmer wrote: *by daily adjusting the vertebrae … I was not only performing a normal obligation, but also a religious duty.*[5] He was convinced that the representative of the "Innate Intelligence" was God within each person and that he had discovered a natural law that pertained to human health in the most general way. At one stage, DD and his son BJ Palmer even toyed with the idea of becoming the founders of a new religion and wrote that *the religion of chiropractic and the religious duty of a chiropractor are one and the same* (see footnote 4). DD declared that he had discovered the answer to the time-worn question, what is life?, and added that chiropractic made *this stage of existence much more efficient in its preparation for the next step—the life beyond* (see footnote 2).

Spinal manipulations or adjustments were not originally meant by Palmer as techniques for treating spinal or musculoskeletal problems; he saw them as a cure for all human illness and stated that *disease is caused by displaced vertebrae or other joints pressing against nerves* (see footnote 4), and that *95% of all diseases are caused by displaced vertebrae, the remainder by subluxations of other joints* (see footnote 2). Early chiropractic pamphlets rarely mentioned back or neck pain, but asserted that, chiropractic could address ailments as diverse as insanity, sexual dysfunction, measles and influenza. Palmer was convinced that he had *created a science of principles that has existed as long as the vertebra* (see footnote 2). He envisioned man as a microcosm of the universe where

---

[5] Palmer DD. *The Chiropractor: The Philosophy and History of Chiropractic Therapy, Care and Diagnostics by its Founder.* lulu.com (2018).

"innate intelligence" determines human health as much as "universal intelligence" governs the cosmos; the discovery of the "innate intelligence" was in his view a discovery of the first order, a *reflection of a critical law that God used to govern natural phenomena* (see footnote 2).

Palmer's gospel spread fast. By 1925, more than 80 chiropractic schools had sprung up in the US. Most were mere "diploma mills" promising an easy way to make money. Chiropractors believed they had established their own form of science, which emphasized observation rather than experimentation, a vitalistic rather than mechanistic philosophy, and a mutually supportive rather than antagonist relationship between science and religion. The gap between conventional medicine and chiropractic soon widened from a fissure into a canyon.

Such rivalry was not confined to conventional medicine, but extended also to osteopaths. DD Palmer taught his students: *don't do anything as an osteopath does* (see footnote 4). Some osteopaths asserted that chiropractic was a bastardized version of osteopathy (see footnote 1). In papers dated 1899 and held at the Palmer College of Chiropractic DD Palmer admitted to have 'borrowed' from osteopathy:

> Some years ago I took an expensive course in Electropathy, Cranial Diagnosis, Hydrotherapy, Facial Diagnosis. Later I took Osteopathy [which] gave me such a measure of confidence as to almost feel it unnecessary to seek other sciences for the mastery of curable disease. Having been assured that the underlying philosophy of chiropractic is the same as that of osteopathy…Chiropractic is osteopathy gone to seed.[6]

In 1924, BJ Palmer introduced the neurocalometer, a heat-sensing instrument purported to detect subluxation.[7] The instrument was advertised as a remarkable innovation with multiple uses and advantages:

The Neurocalometer is a very delicate, sensitive instrument which, when placed upon the spine:

1 *Verifies the proper places for adjustments.*
2 *It measures the specific degree of vertebral pressures upon nerves.*
3 *It measures the specific degree of interference to transmission of mental impulses as a result of vertebral pressure.*
4 *It proves the exact intervertebral foramina that contains bone pressure upon nerves.*

---

[6]Leach, Robert: *The Chiropractic Theories: A Textbook of Scientific Research.* Lippincott, Williams and Wilkins. 2004, p. 15.
[7]Keating JC Jr. Introducing the Neurocalometer: a view from the Fountain Head. *J Can Chiropr Assoc.* 1991;35(3):165–178.

5 *It proves when the pressure has been released upon nerves at a specific place.*
6 *It proves how much pressure was released, if any.*
7 *It verifies the differences between cord pressure or spinal nerve pressure cases.*
8 *It establishes which cases we can take and which we should leave alone.*
9 *It proves by an established record which you can see thereby eliminating all guesswork on diagnoses.*
10 *It establishes, from week to week, whether you are getting well or not.*
11 *It makes possible a material reduction in time necessary to get well, thus making health cheaper….(see footnote 7)*

In reality, the neurocalometer was a useless scam. Yet, BJ forced all his followers to lease it at exorbitant costs. It has been called a model of unethical promotions in health care, (see footnote 7) and divided the Universal Chiropractors' Association thus precipitating the formation of the National Chiropractic Association (NCA), forerunner of today's American Chiropractic Association.

Critical voices were soon raised against chiropractic. In 1924, Henry Louis Mencken (1880–1956) published an essay on chiropractic, many aspects of which are still relevant today[8]:

> This preposterous quackery [chiropractic] flourishes lushly in the back reaches of the Republic, and begins to conquer the less civilized folk of the big cities. As the old-time family doctor dies out in the country towns, with no competent successor willing to take over his dismal business, he is followed by some hearty blacksmith or ice-wagon driver, turned into a chiropractor in six months, often by correspondence… [Chiropractic] pathology is grounded upon the doctrine that all human ills are caused by pressure of misplaced vertebrae upon the nerves which come out of the spinal cord — in other words, that every disease is the result of a pinch. This, plainly enough, is buncombe. The chiropractic therapeutics rest upon the doctrine that the way to get rid of such pinches is to climb upon a table and submit to a heroic pummeling by a retired piano-mover. This, obviously, is buncombe doubly damned…

The first article on chiropractic listed in 'Medline', the world's largest database of medical papers, was published in 1913 in the 'California State Journal of Medicine'[9]:

---

[8] Available at https://www.chirobase.org/12Hx/mencken.html.
[9] "Chiropractic" modesty. *Cal State J Med*. 1913;11(6):213.

Some people are really so terribly modest that it is a mystery how they can live, or even be willing to live, in a world so filled with pushing braggarts and rampant commercialism. For example, note the list of things that E. R. Blanchard D.C., (graduate chiropractor), intimates that he can cure:

"Adhesions, anemia, asthma, appendicitis, blood poison, bronchitis, backache, biliousness, catarrh, constipation, chills and fever, diabetes, dropsy, dizziness, drug and alcohol habits, diarrhoea, deafness, eczema, eye diseases, female diseases, gallstones, gravel, goitre, hay fever, indigestion, lumbago, locomotor ataxia, malaria, nervousness, neuralgia, paralysis, piles, pneumonia, rickets, ruptures, rheumatism, St. Vitus' dance, suppressed or painful menstruation, scrofula, tumors, worms, bed wetting and other child's diseases, leucorrhoea, or whites, stricture, emissions, impotence and many other diseases."

This is almost as long a list as that compiled by the wealthy and admired Law brothers in connection with what they say they can cure with the wonderful Viavi, that prize of all fakes!

Chiropractors' distain for the medical profession is evident in DD Palmer's early texts: *physicians deal with the physical only; chiropractors with both the physical and the spititual* (see footnote 4). The American Medical Association (AMA) had always insisted that all competent healthcare providers must have adequate knowledge of the essential subjects such as anatomy, physiology, pathology, chemistry, and bacteriology. By that token, the AMA claimed, chiropractors were not fit for practice.

Prosecutions against chiropractors for practising medicine without a licence, often instigated by state medical boards, became increasingly common. By 1930, about 15,000 chiropractors had thus been taken to court. In turn, chiropractors started conducting political lobbying to secure licensing statutes. They eventually succeeded in all US states, from Kansas in 1913 to Louisiana in 1974. In turn, chiropractors accused doctors of merely defending their lucrative patch and claimed that orthodox science was morally corrupt and lacked open-mindedness. They attacked the "germo-anti-toxins-vaxiradi-electro-microbioslush death producers" and promised a medicine "destined to the grandest and greatest of this or any age (see footnote 2)." Eventually, the escalating battle against the medical establishment was won in what chiropractors like to call "the trial of the century." In 1987, the U.S. medical establishment were found "guilty of conspiracy against chiropractors".[10]

But such victories came at the price of "taming" and "medicalizing" chiropractic , a process that formed the basis of a conflict within the chiropractic

---

[10]US judge finds medical group conspired against chiropractors, New York Times, 29 August 1987.

profession: the dispute between "mixers" and "straights", a conflict which continues to the present day. Put simply, the "straights" religiously adhere to Palmer's notions of the "innate intelligence" and view subluxation as the sole cause and manipulation as the sole cure of all human disease. They do not mix any non-chiropractic techniques into their therapeutic repertoire, dismiss physical examination (beyond searching for subluxations) and consider medical diagnosis irrelevant for chiropractic. The "mixers" are somewhat more open to science and the advances of conventional medicine, use various treatments other than spinal manipulation, and tend to see themselves as back pain specialists. DD and BJ Palmer warned that the "mixers" were polluting and diluting the sacred philosophy of chiropractic. Much of the endless wrangling within the chiropractic profession during the 20th century was due to this tension. Even today, many straight chiropractors agree that the mixers are a discredit to chiropractic.

The International Chiropractic Association represents the "straights" and the American Chiropractic Association the "mixers." What unites all is a determination to dominate healthcare across the globe. In 2019, the World Federation of Chiropractic published their strategic plan for 2019–2022[11]:

The World Federation of Chiropractic (WFC) envisions a world in which people may enjoy universal access to chiropractic so that populations may thrive and reach their full potential. We exist to support and empower chiropractors and chiropractic associations throughout our 7 world regions to realize this vision by promoting the chiropractic profession and the benefits of the services that chiropractors provide…

In Chap. 3, I will discuss how far this 'universal access' to chiropractic has progressed.

Box 2.1

Mr William Harvey Lillard was cleaner of the Ryan Building where D. D. Palmer's magnetic healing office was located. In 1895, he became Palmer's very first chiropractic patient and thus entered the history books. The foundations of chiropractic are based on this story.

---

[11] Document available at https://www.wfc.org/website/images/wfc/docs/Strategic_Plan_2019-2022/WFC_STRATEGIC_PLAN_2019-2022.pdf.

DEAF SEVENTEEN YEARS.

I was deaf 17 years and I expected to always remain so, for I had doctored a great deal without any benefit. I had long ago made up my mind to not take any more ear treatments, for it did me no good.

Last January Dr. Palmer told me that my deafness came from an injury in my spine. This was new to me; but it is a fact that my back was injured at the time I went deaf. Dr. Palmer treated me on the spine; in two treatments I could hear quite well. That was eight months ago. My hearing remains good.

HARVEY LILLARD,
320 W. Eleventh St., Davenport, Iowa.

[Testimony of Harvey Lillard regarding the events surrounding the first chiropractic adjustment, printed in the January 1897 issue of the Chiropractor]

The nerve supply of the inner ear, the structure that enables us to hear, does not, like most other nerves of our body, run through the spine; it comes directly from the brain: the acoustic nerve is one of the 12 cranial nerves. Therefore, it is not plausible that spinal manipulation might cure deafness. In other words, the story of the 1st chiropractic cure is bogus.

Box 2.2

DD Palmer's description of his cure of a patient with heart disease

Shortly after this relief from deafness, I had a case of heart trouble which was not improving. I examined the spine and found a displaced vertebra pressing against the nerves which innervate the heart. I adjusted the vertebra and gave immediate relief—nothing "accidental" or "crude" about this. Then

I began to reason if two diseases, so dissimilar as deafness heart trouble, came from impingement, a pressure on nerves, were not other disease due to a similar cause, Thus the science (knowledge) and art (adjusting) of Chiropractic were formed at that time. I then began a systematic investigation for the cause of all diseases and have been amply rewarded.

Box 2.3

Milestones in the life of DD Palmer

1845, 7 March: birth in Port Perry, Canada.

1865, 3 April: Palmer family immigrate to the US.

1867: DD Palmer starts as a teacher in Concord, Iowa.

1869, November: DD and his younger brother TJ become beekeepers in Letts, Iowa.

1871, 20 January: DD marries Abba Lord who calls herself a 'psychometrist, clairvoyant physician, soul reader and business medium'.

1872, 6 July: DD publishes an article in the 'Religio-Philosophical Journal' calling himself an 'atheist'.

1872: DD later states that he started his career as a 'healer' during this period.

1873: Abba leaves DD and later becomes a 'homeopathic physician' in Minneapolis.

1876, 7 October: DD marries Louvenia Landers, a widow; they have 4 children together.

1878, 19 April: the Palmer's 5-month old daughter dies.

1878, May: DD is elected president of the 'Western Illinois and Eastern Iowa Society of Bee Keepers'.

1880: DD publishes a pamphlet about spiritualism and refers to himself as a 'spiritualist'.

1881: BJ Palmer is born; he later takes over the chiropractic business and is often referred to as the 'developer of chiropractic'.

1882: DD sells his beekeeping business, moves to What Cheer, Iowa where the rest of his family live.

1883, 30 May: DD opens a grocery store in What Cheer.

1884, 20 November: Louvenia dies of consumption.

1885, February: DD sells his grocery store and 'moves on'.

1885, 25 May: DD marries Martha Henning. The marriage is short-lived; on 8 July of the same year, DD posted a public notice in the 'What Cheer Patriot' disowning her.

1885: DD moves back to Letts where he teaches at the local school.

1886: DD moves to Iola, Kansas where he practices as a magnetic healer and calls himself 'Dr Palmer, healer'.

1886, 3 September: DD advertises his services as a 'vitalist healer' in Burlington, Iowa.

1887, 9 October: DD advertises 'dis-ease is a condition of not ease, lack of ease', a theme that he later uses for chiropractic.

1887, 25 October: one of DD's patients has dies and there is an inquest. The local paper describes DD with the term 'dense ignorance' and the coroner states that 'we censure the so-called doctor, DD Palmer, attending physician, for his lack of treatment and ignorance in the case'. DD leaves Burlington to avoid persecution (a new law requires all healers to register with the state medical board. DD does not have such a registration).

1887: DD moves to Davenport and advertises: DD Palmer, cures without medicine...'

1888, 6 November: DD marries Villa; they stay together until her death in 1905.

1894: DD publishes his views on smallpox vaccination: '...the monstrous delusion ... fastened on us by the medical profession, enforced by the state boards, and supported by the mass of unthinking people ...'

1894: DD publishes his views on 'greedy doctors' and the 'medical monopoly'.

1895, January: DD starts a business selling goldfish.

1895, 18 September: DD administers the 1st spinal manipulation to Harvey Lillard (DD later seems confused about this date stating that this 'was done about Dec. 1st, 1895').

1896, 14 January is the date when, according to DD, chiropractic received its name with the help of Reverent Weed.

1896: DD publishes an article in 'The Magnetic' stating 'the magnetic cure: how to get well and keep well without using poisonous drugs'.

1896: DD publishes his theory that bacteria cannot grow on healthy tissue; keeping tissue healthy is therefore the best prevention against infections; and this is best achieved by magnetic healing.

1896: DD claimed that 4 years earlier, in 1892, he had discovered the magnetic cure for cancer; it involved freeing the stomach and spleen of poisons.

1896: DD formulates his concept of treating the root cause of any disease.

1896, 10 July: DD, his wife and his brother turn the 'Palmer School of Magnetic Cure' in Davenport into an officially registered corporation.

1897: DD defines chiropractic as 'a science of healing without drugs'.

1898: DD opens his first school of chiropractic in Davenport, the 'Palmer School of Chiropractic' which has survived to the present day.

1902, 27 April: DD first uses the term 'subluxation' in a letter to his son BJ ('... where you find the greatest heat, there you will find the subluxation causing the inflammation which produces the fever...').

1902: DD leaves suddenly for California, apparently to open a West Coast branch of the Palmer School; he stays for about two years and then returns to Davenport leaving behind substantial debts.

1902, 6 September: DD is arrested in Pasadena when a patient suffering from consumption dies after DD's second adjustment; in October, the charges were dropped because of a technicality.

1903: DD opens the 'Palmer Chiropractic School in Santa Barbara, California, together with his former student Oakley Smith.

1903 DD is charged with practising medicine without licence but, before the case comes to trial, DD moves to Chicago where he opens a school together two other chiropractors (Smith and Paxson); however, the project fails.

1903, 30 April: DD is back in Davenport for the wedding of BJ with Mabel.

1904, December: DD starts his new journal 'The Chiropractor' which survives until 1961. DD's very first article is entitled '17 Years of Practice'.

1905: DD's former students Langworthy and Smith accuse DD of stealing the concepts of chiropractic from the Bohemian bonesetters of Iowa.

1905, 9 November: DD's wife Villa overdoses on morphine and dies; the coroner is unable to tell whether she committed suicide or intended it for pain relief.

1906, 11 January: DD marries Mary Hunter, apparently his first love from Letts.

1906, 26 March: DD is again on trial for practising medicine without a licence. He is found guilty the next day. The penalty is US$ 350 or 105 days in jail. DD choses jail. However, his new wife, Mary, bails him out after 23 days.

1906: DD sells his share in the chiropractic business to BJ and moves to Medford Oklahoma. The reasons for this split are said to be personal, financial and professional.

1906, 4 June: in a letter to John Howard, DD accuses BJ of dishonesty and of running the school badly.

1906: BJ and DD publish their opus maximus 'Science of Chiropractic'; DD claims that most of the chapters were written by him.

1907, January: DD opens another grocery store.

1908: together with a colleague, DD opens the 'Palmer-Gregory Chiropractic College'; it lasts only 9 weeks. DD left because he discovered that Alva Gregory, a medical doctor, was teaching medical concepts.

1908, 9 November: DD opens the 'Palmer College of Chiropractic' in Portland, Oregon.

1908, December: DD starts a new journal, 'The Chiropractor's Adjuster'; many of his articles focus on criticising BJ. The journal only seems to have survived until 1910.

1910, December: DD publishes his book 'The Chiropractor's Adjuster'.

1911: DD toys with the idea of turning chiropractic into a religion, as this would avoid chiropractors being sued for practising medicine without a license, he states.

1913: DD visits Davenport for the 'Lyceum Parade' where he is injured. Mary accuses BJ of striking his father with his car and thus indirectly causing his death, a version of events which is disputed.

1913, September: DD is back in California and writes to JB Olson that he gave 22 lectures in Davenport. DD also reports: '... On the return I cured a man of sun stroke by one thrust on the 5th dorsal. That is what I call definitive, specific, scientific chiropractic...'.

1913, 20 October: DD dies; the official cause of death is typhoid fever, a condition he had repeatedly claimed to be curable by a single spinal adjustment.

1914: DD Palmer's book 'The Chiropractor' is published.

Box 2.4

Definitions of chiropractic

During the last 120 years, many different definitions of chiropractic (see footnote 1) have emerged.

Chiropractic is:

- the art of adjusting by hand all subluxations of the three hundred articulations of the human skeletal frame, more especially the 52 articulations of the spinal column, for the purpose of freeing impinged nerves, as they emanate thru the intervertebral foramina, causing abnormal function, in excess or not, named disease (DD Palmer, BJ Palmer 1906).
- a form of alternative medicine mostly concerned with the diagnosis and treatment of mechanical disorders of the musculoskeletal system, especially the spine (Wikipedia).
- a system of therapy which holds that disease results from a lack of normal nerve function and which employs manipulation and specific adjustment of body structures such as the spinal column (Merriam Webster).
- a method of treatment that manipulates body structures (especially the spine) to relieve low back pain or even headache or high blood pressure (sensagent).
- a system of treating bodily disorders by manipulation of the spine and other parts, based on the belief that the cause is the abnormal functioning of a nerve (Collins English Dictionary).
- a science of healing without drugs (DD Palmer).
- a natural form of health care that uses spinal adjustments to correct these misalignments and restore proper function to the nervous system, helping your body to heal naturally (Palmer College of Chiropractic).
- a non-invasive, hands-on health care discipline that focuses on the musculoskeletal system (Ontario Chiropractic Association).
- a health care profession that focuses on disorders of the musculoskeletal system and the nervous system, and the effects of these disorders on general health (American Chiropractic Association).
- a healthcare discipline that emphasizes the inherent recuperative power of the body to heal itself without the use of drugs or surgery. The practice of chiropractic focuses on the relationship between structure (primarily the spine) and function (as coordinated by the nervous system) and how that relationship affects the preservation and restoration of health. In addition, doctors of chiropractic recognize the value and responsibility of working in cooperation with other health care practitioners when in the best interest of the patient. (Association of Chiropractic Colleges).
- a health care profession concerned with the diagnosis, treatment and prevention of disorders of the neuromusculoskeletal system and the effects of these disorders on general health. There is an emphasis on manual techniques, including joint adjustment and/or manipulation with a particular focus on subluxations (WHO).
- a separate and distinct profession dedicated to the detection and correction of vertebral subluxation for the better expression of life (International Federation of Chiropractors and Organizations).
- a health profession concerned with the diagnosis, treatment and prevention of mechanical disorders of the musculoskeletal system, and the effects of these disorders on the function of the nervous system and general health.

There is an emphasis on manual treatments including spinal adjustment and other joint and soft-tissue manipulation (World Federation of Chiropractic).
- a treatment where a practitioner called a chiropractor uses their hands to help relieve problems with the bones, muscles and joints (NHS England).

3

Current Popularity of Chiropractic

We envision a world where people of all ages, in all countries, can access the benefits of chiropractic.[1] (The World Federation of Chiropractic, 2019)

Chiropractic is hugely popular—at least this is what we are often told, usually by interested parties. But how popular is it really? The answer is, as we shall see, that its popularity is variable and usually modest.

The aim of this analysis[2] was to study the prevalence, patterns, and predictors of chiropractic utilization in the US general population. Cross-sectional data from the 2012 National Health Interview Survey (n = 34,525) were analysed. Lifetime and 12-month prevalence of chiropractic use were 24.0% and 8%, respectively. Back pain (63.0%) and neck pain (30.2%) were the two conditions for which chiropractic was used most.

Our own summary of all representative surveys of the population suggested that the one-year prevalence of chiropractic usage varied between 7% in the US (1997) and 33% in Scotland (1996).[3] The findings of a large survey of

---

[1] https://www.wfc.org/website/index.php?option=com_content&view=article&id=533:wfc-releases-new-guiding-principles-document&catid=56:news--publications&Itemid=27&lang=en&fbclid=IwAR26GqFjENyJEfGD9_Pseoo-9jcnX8UlrFKXiyFuIOEUWPbqThO5IL_yvew.

[2] Adams J, Peng W, Cramer H, et al. The Prevalence, Patterns, and Predictors of Chiropractic Use Among US Adults: Results From the 2012 National Health Interview Survey. *Spine (Phila Pa 1976)*. 2017; 42(23):1810–1816. https://doi.org/10.1097/BRS.0000000000002218.

[3] Ernst E. Prevalence of use of complementary/alternative medicine: a systematic review, Bulletin of the World Health Organization, 2000, 78(2). https://www.ncbi.nlm.nih.gov/pmc/articles/PMC2560678/pdf/10743298.pdf.

© Springer Nature Switzerland AG 2020
E. Ernst, *Chiropractic*,
https://doi.org/10.1007/978-3-030-53118-8_3

a random sample of the UK general population implied that, in 1999, only 3% used chiropractic services.[4]

Some of the basic facts about chiropractic include the following:

- There are currently around 15 000 chiropractors practising in the US[5] (significantly more than anywhere else, except for Australia (Box 3.1)).
- The average yearly income of a US chiropractor is around US$ 60 000 (see footnote 3).
- US chiropractors provide ~20 million treatments per year under Medicare (see footnote 3).
- Overall spending for chiropractic services was estimated to be US$ $12.5 billion in 2015 (see footnote 3).
- The average length of a chiropractic visit is about 22 min.[6]
- The cost for one treatment varies between US$ 30 and 200 (see footnote 4).
- About 27% of US physicians recommend chiropractic or osteopathic spinal manipulations to their patients.[7]
- The global chiropractic care market was US$ 577 million in 2018 and is estimated to reach US$ 847 million by 2025.[8]

A global overview, published in 2017,[9] summarised the situation regarding chiropractic. The average use of chiropractic within a year was stable between 1980 and 2015 and amounted to 9% which is close to the US data cited above. The commonest reasons for people attending chiropractic care were low back pain (49.7%), neck pain (22.5%), as well as problems in arms and legs (10.0%). The treatments provided by chiropractors were spinal manipulations (79.3%), soft-tissue therapy (35.1%), and patient education (31.3%). Three percent of the general population sought chiropractic care for non-musculoskeletal conditions. The reasons for children to have chiropractic treatments seem particularly noteworthy:

---

[4] Ernst E, White A. The BBC survey of complementary medicine use in the UK. *Complement Ther Med*. 2000; 8(1):32–36.

[5] Long PH. Chiropractic Abuse: An Insider's Lament. American Council on Science and Health (2013).

[6] Yeh GY, Phillips RS, Davis RB, Eisenberg DM, Cherkin DC. Visit time as a framework for reimbursement: time spent with chiropractors and acupuncturists. *Altern Ther Health Med*. 2003; 9(5):88–94.

[7] Stussman BJ, Nahin RR, Barnes PM, and Ward BW. U.S. Physician Recommendations to Their Patients About the Use of Complementary Health Approaches. The Journal of Alternative and Complementary Medicine. Jan 2020, 25–33; https://doi.org/10.1089/acm.2019.0303.

[8] https://www.openpr.com/news/1933382/the-global-chiropractic-care-market-which-was-usd-576-88-million.

[9] Beliveau PJH, Wong JJ, Sutton DA, et al. The chiropractic profession: a scoping review of utilization rates, reasons for seeking care, patient profiles, and care provided. *Chiropr Man Therap* **25**, 35 (2017). https://doi.org/10.1186/s12998-017-0165-8.

- 7% for infections,
- 5% for asthma,
- 5% for stomach problems,
- 5% for wellness/maintenance.

As we shall see in Chap. 12, none of these indications are even remotely evidence-based. Remarkably, 35% of chiropractors employed X-ray diagnostics (Chap. 15), and only 31% conducted a medical history of their patients. The latter finding might be less shocking, if we consider DD Palmer's dictum: *disease is caused by displaced vertebrae or other joints pressing against nerves.*[10]

In 2016, the World Federation of Chiropractic Disability and Rehabilitation Committee published a survey of all 193 United Nation member countries and seven dependencies describing the global chiropractic workforce.[11] Information was available from 90 countries. The total number of chiropractors worldwide was 103,469 (in comparison, there were 450,000 physical therapists worldwide). The findings show that:

- Chiropractic education was offered in 48 institutions in 19 countries.
- Direct access to chiropractic services was available in 81 countries.
- Chiropractic services were partially or fully covered by government and/or private health schemes in 46 countries.
- Chiropractic was legally recognized in 68 countries.
- Chiropractic was explicitly illegal in 12 countries.
- Chiropractors' scope of practice was governed by legislation or regulation in just 26 countries.
- The professional title of chiropractic was protected by legislation in 39 countries.
- In 43 countries, chiropractors were permitted to own X-ray machines, to operate them, and to prescribe X-rays.
- In 22 countries, they were permitted to prescribe advanced imaging (MRI or CT).
- In 34 countries, chiropractors were permitted to own and operate diagnostic ultrasound equipment.
- Full or limited rights to the prescription of pharmaceutical medication were granted in 9 countries.
- Authorization of sick leave by chiropractors was permitted in 20 countries.

---

[10]Palmer DD. *The Chiropractor: The Philosophy and History of Chiropractic Therapy, Care and Diagnostics by its Founder.* lulu.com (2018).

[11]Stochkendahl MJ, Rezai M, Torres P, et al. The chiropractic workforce: a global review. *Chiropr Man Therap* **27**, 36 (2019). https://doi.org/10.1186/s12998-019-0255-x.

- The care of children was subject to specific regulations and/or statutory restrictions in 57 countries.

The countries in which chiropractic is explicitly illegal were:

- Egypt,
- Argentina,
- Columbia,
- Austria,
- Estonia,
- Greece,
- Hungary,
- Lebanon,
- Republic of Korea,
- Taiwan,
- Turkey,
- Ukraine.

The educational requirements for chiropractors obviously differ from country to country (Boxes 3.2 and 3.3). An Australian survey from 2016 monitored the views of general practitioners towards chiropractors.[12] Its findings suggested that 70% of all general practitioners believed chiropractic education was not evidence-based, that 60% of them had never referred a patient to a chiropractor, and that they would not want to co-manage patients with a chiropractor. The authors concluded that *attitudes may have become less favourable with a growing intolerance towards chiropractors.*

Our own survey of UK general practitioners suggested that, in 1999, 7% of them referred patients to chiropractors, and 11% of them endorsed chiropractic. By 2010, these two percentage figures had fallen to 4 and 7%, respectively[13] indicating a significant loss of acceptance.

A Canadian survey described the profiles of chiropractors' practice, the nature of the care provided to patients and the extent of interprofessional

---

[12] Engel RM, Beirman R, and Grace S. An indication of current views of Australian general practitioners towards chiropractic and osteopathy: a cross-sectional study. *Chiropr Man Therap* **24**, 37 (2016). https://doi.org/10.1186/s12998-016-0119-6.

[13] Perry R, Dowrick C, Ernst E. Complementary medicine and general practice in an urban setting: a decade on. *Prim Health Care Res Dev.* 2014;15(3):262–267. https://doi.org/10.1017/S14634236130 00182.

collaborations.[14] The researchers recruited 120 chiropractors who were in active practice in 2015. Each chiropractor recorded information for up to 100 consecutive patient encounters. Thus, data on 3523 chiropractor-patient encounters became available. The findings suggest that the typical patient is female (59% of encounters) and between 45 and 64 years old (43%). Only 7% were referred by physicians and 68% of patients paid out of pocket or claimed extended health insurance for care. The most common diagnoses were back (49%) and neck (15%) problems. The most common treatments included spinal manipulation (72%), soft tissue therapy (70%) and mobilisation (35%). Chiropractors also employed various other therapies none of which can be categorised as solidly based on evidence:

- ultrasound 3%,
- acupuncture 3%,
- traction 1%,
- interferential therapy 3%,
- soft laser therapy 3% (Chap. 6).

Remarkably, 54% of all patients reported being in "excellent/very good overall health". This, of course, begs the question, why then did they bother to consult a chiropractor? One answer might be disease prevention, a subject discussed in Chap. 12.

A 2012 report by the WHO identified four main challenges chiropractors might face in the future[15]:

1. Lack of Funding for Education and Research. There is still only modest public funding for chiropractic education and research, most of this activity being funded by the profession itself. With respect to education, results are high student debt for those graduating from private colleges, small class sizes in government-funded universities, and delays in the opening of new programs. With respect to research, there is increasing competition for research grants in most countries.
2. Financial Barriers to Patient Access. In most countries patients continue to experience financial barriers to access when choosing to consult a chiropractor. This may be because of exclusion from government and private

---

[14]Mior S, Wong J, Sutton D, et al. Understanding patient profiles and characteristics of current chiropractic practice: a cross-sectional Ontario Chiropractic Observation and Analysis STudy (O-COAST), *BMJ Open* 2019; **9:**e029851. https://doi.org/10.1136/bmjopen-2019-029851.

[15]The Current Status of the Chiropractic Profession. Report to the World Health Organization from the World Federation of Chiropractic, December, 2012. Document available from https://www.wfc.org/website/images/wfc/WHO_Submission-Final_Jan2013.pdf.

health plans or, where most plans include some coverage for chiropractic services as in North America, because of co-payments and limits that are more restrictive than for other providers. Achieving parity in this area is a continuing challenge.

3. More Policy Input and Research. Chiropractors are beginning to be appointed to senior advisory and policy positions in-health systems in some countries, but a current challenge is to build successfully upon recent gains contributing to health policy, public health and research. This will provide the necessary foundation for more complete integration in health systems in the management of spinal health and wellness. Research priorities include: Back Pain—e.g. identifying subgroups that respond to care; back pain and disability as a public health issue in pediatric, working age, and senior populations. Service Delivery—e.g. interdisciplinary spine care guidelines and pathways; patient demographics; models of care. Value of Preventive Care—e.g. measuring well-being and prevention of injuries and disability in specific (e.g. office workers, manual workers, elite athletes) and general populations. Cost Effectiveness and Value in Addressing Health System Problems—e.g. ability to reduce cost of spine care and wait times in primary care, for specialist services and in hospital emergency departments.

4. Lack of Growth and Availability. Chiropractic services are well-accepted but difficult to locate in many countries because of small numbers of chiropractors in practice. There is a need for new chiropractic educational programs in all world regions other than in North America.

In the following chapters, I will address some of these challenges in more detail.

Box 3.1 Estimated number of chiropractors practising in countries other than the US

Australia 2500
Brazil 700
Denmark 550
France 450
Italy 400
Japan 400
Netherlands 400
New Zealand 400
Norway 600
UK 2000
US 1500
South Afrika 400

Spain 300
Sweden 400
Switzerland 300.

Box 3.2 Educational requirements and licensure of US chiropractors

Chiropractors must:

- earn a Doctor of Chiropractic (D.C.) degree,
- pass the National Board of Chiropractic Examiners exam,
- obtain a state license and engage in continuing education.

A D.C. programmes usually take 4 years.

Students must have had at least 3 years of undergraduate education.

Chiropractic schools must be accredited by the Council on Chiropractic Education.

Classes cover basic sciences, such as anatomy and physiology, and supervised clinical experience.

Box 3.3 Educational requirements for UK chiropractors

Chiropractors must be registered with the General Chiropractic Council (GCC), the statutory body responsible for regulating the chiropractic profession in the UK.

Chiropractors must complete a GCC-recognised degree from one of the following institutions:

- AECC University College in Bournemouth offers a four to five-year, depending on your year of entry, full time Master of Chiropractic (MChiro) Hons.
- London South Bank University offers a Chiropractic Integrated Masters (MChiro), a four-year full-time course in their Southwark Campus.
- McTimoney College of Chiropractic offers a four and five-year integrated MChiro validated by BPP University. The five-year programme is aimed at students wanting to continue working while studying. There are learning centres in Abingdon, Oxfordshire, and in Manchester.
- The University of South Wales offers a four-year integrated MChiro.

Courses offer basic sciences, such as anatomy and physiology, practical training in adjustment and supervised clinical training, research methods, business and professional and ethical practice.

Entry requirements include GCSEs in mathematics and English and three A-levels, or other relevant higher qualifications.

4

Subluxation

Subluxation is … a displacement of two or more bones whose articular surfaces have lost, wholly or in part, their natural connection.
—D. D. Palmer (1910)

The definition of 'subluxation' as used by chiropractors differs from that in conventional medicine where it describes a partial dislocation of the bony surfaces of a joint readily visible via an X-ray. Crucially, a subluxation, as understood in conventional medicine, is not the cause of disease. Spinal subluxations, according to medical terminology, are possible only if anatomical structures are seriously disrupted.

Subluxation, as chiropractors understand the term, has been central to chiropractic from its very beginning (Box 4.1).[1] Despite its central role in chiropractic, its definition is far from clear and has changed significantly over time (Box 4.2).

Palmer was extremely vague about most of his ideas. Yet, he remained steadfast about his claims that 95% of all diseases were due to subluxations of the spine, that subluxations hindered the flow of the 'innate intelligence' which controlled the vital functions of the body. Innate intelligence or 'inate', he believed, operated through the nerves, and subluxated vertebra caused pinched nerves, which in turn blocked the flow of the innate and thus led

[1] Senzon SA. The Chiropractic Vertebral Subluxation Part 2: The Earliest Subluxation Theories From 1902 to 1907. *J Chiropr Humanit*. 2019; 25:22–35. Published 2019 April 6. doi:https://doi.org/10.1016/j.echu.2018.10.009.

© Springer Nature Switzerland AG 2020
E. Ernst, *Chiropractic*,
https://doi.org/10.1007/978-3-030-53118-8_4

to abnormal function of our organs.[1] For Palmer and his followers, subluxation is the sole or at least the main cause of all diseases (or dis-eases, as Palmer preferred).

Palmer's theory has always been controversial, not least because it contradicted elementary anatomical facts and established medical knowledge. While spinal nerves can, of course, be irritated by intervertebral discs, they cannot be impinged by displaced bones unless there is a serious pathology, such as a vertebral fracture.[2]

Throughout DD Palmer's 1910 book entitled '*Textbook of the Science, Art, and Philosophy of Chiropractic*' (https://www.worldcat.org/title/textbook-of-the-science-art-and-philosophy-of-chiropractic-for-students-and-practitioners/oclc/17205743), he made it clear that the identity of chiropractic depends on his metaphysical concept of vertebral subluxation. In 1909, Palmer's son, BJ, stated that "*Chiropractors have found in every disease that is supposed to be contagious, a cause in the spine. In the spinal column we will find a subluxation that corresponds to every type of disease. If we had one hundred cases of small-pox, I can prove to you where, in one, you will find a subluxation and you will find the same conditions in the other ninety-nine. I adjust one and return his functions to normal.... There is no contagious disease.... There is no infection.... There is a cause internal to man that makes of his body in a certain spot, more or less a breeding ground [for microbes]. It is a place where they can multiply, propagate, and then because they become so many they are classed as a cause.*"[3]

The concept of subluxation is so central to chiropractic because:

- it gives the profession its uniqueness,
- it legitimizes chiropractic,
- it is consistent with the intentions of chiropractic's creator.[4]

Without subluxation, there is no chiropractic spinal manipulation, and without spinal manipulation, there is no chiropractic (Box 4.3). This logic goes a long way towards explaining why even today, in the face of overwhelming evidence, many chiropractors are ferociously reluctant to abandon the concept of subluxation and defend it at all cost. It also explains why the 'Practice Analysis of Chiropractic 2020' by the 'National Board of

---

[2]Crelin ES. A scientific test of the chiropractic theory. *Am Sci*. 1973; 61(5):574–580.
[3]Campbell JB, Busse JW, Injeyan HS. Chiropractors and Vaccination: A Historical Perspective. Pediatrics Apr 2000, 105(4) e43; https://doi.org/10.1542/peds.105.4.e43.
[4]Hart J. Analysis and Adjustment of Vertebral Subluxation as a Separate and Distinct Identity for the Chiropractic Profession: A Commentary. *J Chiropr Humanit*. 2016;23(1):46–52. Published 2016 Oct 25. https://doi.org/10.1016/j.echu.2016.09.002.

Chiropractic Examiners' shows that 80% of US chiropractors diagnose a subluxation in their patients.[5]

For many years, chiropractors have insisted that subluxations are detectable via X-ray diagnostics (this is one reason why they grossly and dangerously overuse X-rays).[6] When both the metaphysical and the 'bone out of place' theories of subluxation had become untenable, they worked towards and eventually succeeded in changing the definition of subluxation into a more functional one. In 1996, the 'Association of Chiropractic Colleges' (ACC) published a consensus definition of subluxation, later also adopted by the World Federation of Chiropractic, to accommodate (most of) the diverse views on the subject, even Palmer's notion that subluxation is the cause of disease[7]:

> Chiropractic is concerned with the preservation and restoration of health and focuses particular attention on the subluxation. A subluxation is a complex of functional and/or structural and/or pathological articular changes that compromise neural integrity and may influence organ system function and general health. A subluxation is evaluated, diagnosed, and managed through the use of chiropractic procedures based on the best available rational and empirical evidence.

Animal models were occasionally used to support the concept, and some research seemed to suggest that subluxations modulate activity in afferent nerves. However, it remained unclear whether these afferent nerves are modulated during normal day-to-day activities of living and, if so, what segmental or whole-body reflex effects they may have.[8] Crucially, Palmer's theory that subluxations influence organ function and general health of humans has never been backed up by sound evidence.

In 2012, the chiropractor C.N.R. Henderson agreed that the vertebral subluxation has been the historical 'raison d'etre' for spinal manipulation and stated that vertebral subluxation is a biomechanical spine derangement thought to produce clinically significant effects by disturbing neurological

---

[5]Document available at https://mynbce.org/wp-content/uploads/2020/02/Practice-Analysis-of-Chiropractic-2020-3.pdf.

[6]Ernst E. Chiropractors' use of X-rays. *Br J Radiol.* 1998; 71(843):249–251. https://doi.org/10.1259/bjr.71.843.9616232.

[7]Senzon SA. The Chiropractic Vertebral Subluxation Part 10: Integrative and Critical Literature From 1996 and 1997. Journal of Chiropractic Humanities, Vol 25, December 2018, pp 146–168.

[8]Bolton PS. Reflex effects of vertebral subluxations: the peripheral nervous system. An update. *J Manipulative Physiol Ther.* 2000; 23(2):101–103. https://doi.org/10.1016/s0161-4754(00)90075-7.

function.[9] For many chiropractors subluxation had by then quite simply become the abnormality they adjust; negating its existence would thus be tantamount to negating usefulness of chiropractic.

But not every chiropractor agreed, and the long-standing debate regarding the chiropractic subluxation continued to create substantial controversy within the profession. The subluxation construct, some chiropractors wryly claimed, is similar to the Santa Claus construct: they both are myths that survive because of their social utility.[10] Others argued:

> the situation within the chiropractic profession corresponds very much to that of an unhappy couple that stays together for reasons that are unconnected with love or even mutual respect. We also contend that the profession could be conceptualised as existing on a spectrum with the 'evidence-friendly' and the 'traditional' groups inhabiting the end points, with the majority of chiropractors in the middle. This middle group does not appear to be greatly concerned with either faction and seems comfortable taking an approach of 'you never know who and what will respond to spinal manipulation'. We believe that this 'silent majority' makes it possible for groups of chiropractors to practice outside the logical framework of today's scientific concepts.[11]

Leboef-Yde et al. went further, providing the following examples of the subluxation problem as seen from the perspective of evidence-friendly chiropractors:

- In Canada, vitalist practitioners have been shown to be more likely to have anti-vaccination beliefs, and their attitudes about radiographic imaging are inconsistent with current evidence/guideline-based care. As such, vitalistic providers were less likely to receive referrals from other health providers.
- In Florida, U.S.A., attempts to establish a university-based education in chiropractic were stopped in 2005 because of opposition and lobbying from the traditional group.
- In 2009 in the UK, a systematic survey of chiropractic websites was done by a group motivated by displeasure at unsupported claims of chiropractors, and formal complaints were lodged with the General Chiropractic Council. Although most chiropractors were found not guilty, thousands of

---

[9]Henderson CN. The basis for spinal manipulation: chiropractic perspective of indications and theory. *J Electromyogr Kinesiol*. 2012; 22(5):632–642. https://doi.org/10.1016/j.jelekin.2012.03.008.

[10]Good CJ. The great subluxation debate: a centrist's perspective. *J Chiropr Humanit*. 2010; 17(1):33–39. https://doi.org/10.1016/j.echu.2010.07.002.

[11]Leboeuf-Yde C, Innes SI, Young KJ, et al. Chiropractic, one big unhappy family: better together or apart? *Chiropr Man Therap* **27**, 4 (2019). https://doi.org/10.1186/s12998-018-0221-z.

work-hours and much stress was caused. The content of these web sites was subsequently changed.

- In 2012, the treatment of children based on traditional chiropractic 'diagnoses' at the student chiropractic clinics at the Royal Melbourne Institute of Technology in Australia, a university-based chiropractic course, at the time led by a well-known traditionally-oriented chiropractor, brought down both fury and ridicule on chiropractic. It also resulted in a new movement called 'Friends of Science', who wage war on university education involving non-evidence-based alternative medicine, notably chiropractic. This severely threatened at least two chiropractic undergraduate courses.
- In 2013, attempts to establish a university-based education in chiropractic in Sweden were stopped following a debate that exposed unsupported claims on the websites of some chiropractors.

Straight chiropractors, in turn, feel aggrieved by their 'evidence-friendly' colleagues, because 'real' chiropractic is being denigrated or squandered. According to Leboef-Yde et al., this has two main reasons:

- Evidence-friendly chiropractors are seen as unfaithful to the traditional tenets of chiropractic (i.e. subluxation as a basis for changes in health).
- Evidence-friendly chiropractors side with the 'enemy', i.e. medical doctors, research scientists, sceptics, etc.

Critics, such as Keating et al., list three crucial issues with subluxation[12]:

1. The hypothesis that subluxation is some "complex of functional and/or structural and/or pathological articular changes that compromise neural integrity" is offered without qualification, that is, without mention of the tentative, largely untested quality of this claim.
2. The dogmatism of the ACC's unsubstantiated claim that subluxations "may influence organ system function and general health" is not spared by the qualifier "may." The phrase could mean that subluxations influence "organ system function and general health" in some but not all cases, or that subluxation may not have any health consequences. Although the latter interpretation is tantamount to acknowledging the hypothetical status of subluxation's putative effects, this meaning seems unlikely in

---

[12] Keating JC Jr, Charlton KH, Grod JP, Perle SM, Sikorski D, Winterstein JF. Subluxation: dogma or science?. *Chiropr Osteopat*. 2005; 13:17. Published 2005 Aug 10. https://doi.org/10.1186/1746-1340-13-17.

light of the ACC's statement that chiropractic addresses the "preservation and restoration of health" through its focus on subluxation. Both interpretations beg the scientific questions: do subluxation and its correction "influence organ system function and general health"?

3. The ACC claims that chiropractors use the "best available rational and empirical evidence" to detect and correct subluxations. This strikes us as pseudoscience, since the ACC does not offer any evidence for the assertions they make, and since the sum of all the evidence that we are aware of does not permit a conclusion about the clinical meaningfulness of subluxation. To the best of our knowledge, the available literature does not point to any preferred method of subluxation detection and correction, nor to any clinically practical method of quantifying compromised "neural integrity," nor to any health benefit likely to result from subluxation correction.

As early as 1968, the US Department of Health, Education and Welfare published robust assessments of the evidence concluding that there is no evidence that subluxation, if it exists, is a significant factor in disease processes.[13] Several undercover investigations have shown that chiropractors diagnose subluxations in every and even in entirely healthy volunteers, while failing to agree on the site of the alleged abnormality.[14] In other words, as a diagnosis, subluxation is an entirely useless entity; it serves one purpose only: it keeps chiropractors in business.

In 2005 Keating went further, arguing that the dogma of subluxation is perhaps the greatest single barrier to professional development for chiropractors. It skews the practice of the art in directions that bring ridicule from the scientific community and uncertainty among the public. Failure to challenge the subluxation dogma perpetuates a marketing tradition that inevitably prompts charges of quackery. Subluxation dogma leads to legal and political strategies that may amount to a house of cards and warp the profession's sense of self and of mission. Commitment to this dogma undermines the motivation for scientific investigation of subluxation as hypothesis, and so perpetuates the cycle.[15]

Another careful analysis of the concept concluded that *there is a significant lack of evidence in the literature to fulfil Hill's criteria of causation with*

---

[13]Long PH. Chiropractic Abuse: An Insider's Lament. American Council on Science and Health (2013).

[14]https://www.quackwatch.org/01QuackeryRelatedTopics/chiroinv.html.

[15]Keating JC Jr, Charlton KH, Grod JP, Perle SM, Sikorski D, Winterstein JF. Subluxation: dogma or science?. *Chiropr Osteopat*. 2005; 13:17. Published 2005 Aug 10. https://doi.org/10.1186/1746-1340-13-17.

regards to the chiropractic subluxation. No supportive evidence is found for the chiropractic subluxation being associated with any disease process or of creating suboptimal health conditions requiring intervention. Regardless of popular appeal this leaves the subluxation construct in the realm of unsupported speculation. This lack of supportive evidence suggests the subluxation construct has no valid clinical applicability.[16]

In view of such evidence, some chiropractors are keen to reform their profession and to abandon the idea of subluxation altogether. A position paper[17] by a group of six European chiropractic programmes, for instance, stated:

> The teaching of vertebral subluxation complex as a vitalistic construct that claims that it is the cause of disease is unsupported by evidence. Its inclusion in a modern chiropractic curriculum in anything other than an historical context is therefore inappropriate and unnecessary.

In view of these problems, some chiropractic schools have recently started to teach subluxation merely as an historical concept. Yet, despite the lack of validation of subluxation, many chiropractors remain stubbornly unwilling to abandon the concept, and the multiple efforts of vitalistic chiropractors to keep the myth of subluxation alive remain successful. Essentially, they argue that subluxation has to exist because it is vital for their profession: *both clinically and legally there has to be an entity that practitioners identify.*[18]

A 2018 survey determined how many chiropractic institutions worldwide still use the term in their curricula.[19] Forty-six chiropractic programmes (18 from US and 28 non-US) participated. The term subluxation was found in all but two US course catalogues. Remarkably, between 2011 and 2017, the use of subluxation in US courses even increased. Similarly, a survey of 7455 US students of chiropractic showed that 61% of them agreed or strongly

---

[16]Mirtz TA, Morgan L, Wyatt LH, et al. An epidemiological examination of the subluxation construct using Hill's criteria of causation. *Chiropr Man Therap* **17**, 13 (2009). https://doi.org/10.1186/1746-1340-17-13.

[17]http://chiropractic.prosepoint.net/163929.

[18]Rome PL, Waterhouse JD. Evidence Informed Vertebral Subluxation—A Diagnostic and Clinical Imperative. Journal of Philosophy, Principles & Practice of Chiropractic, December 27, 2019, pp 12–34.

[19]Funk MF, Frisina-Deyo AJ, Mirtz TA, et al. The prevalence of the term subluxation in chiropractic degree program curricula throughout the world. *Chiropr Man Therap* **26**, 24 (2018). https://doi.org/10.1186/s12998-018-0191-1.

agreed that the emphasis of chiropractic intervention is to eliminate vertebral subluxations/vertebral subluxation complexes.[20]

Even though chiropractic subluxation is at the heart of chiropractic, its definition remains nebulous and its very existence seems doubtful. But doubt is not what chiropractors want. Without subluxation, spinal manipulation seems questionable—and this will be the theme of the next chapter.

Box 4.1 Early uses of the term subluxation[1]

Chiropractor	Term	First known use	Publication	Quote
D.D. Palmer	Luxation	1900	The Chiropractic	Childbed fever is always caused by a lumber luxation during childbirth
O.G. Smith	Sub-luxations	February 4, 1902	Clarinda Herald	Please bear in mind—do not ever forgot: (a) Sub-luxations of vertebrae and other boned to occur. (b) That all diseases are caused by a pressure upon nerves or blood vessels. (c) And hence, that all diseases (nearly all) can be cured by REPLACING THE MISPLACED STRUCTURES, AND RELIEVING THE PRESSURE ON THE NERVES AND BLOOD VESSELS

(continued)

---

[20]Gliedt JA, Hawk C, Anderson M, et al. Chiropractic identity, role and future: a survey of North American chiropractic students. *Chiropr Man Therap.* 2015; 23(1):4. Published 2015 Feb 2. https://doi.org/10.1186/s12998-014-0048-1.

(continued)

Chiropractor	Term	First known use	Publication	Quote
D.D. Palmer	Sub-luxations	April 27, 1902	Letter to B.J. Palmer	Where you find the greatest heat, there you will find the sub-luxation causing the inflammation
B.J. Palmer	Sub-luxations	August 15, 1902	Davenport Times	Chiropractic is the SCIENCE of replacing these sub-luxations (misplacements), thereby releasing pressures, consequently we remove the causes, and do not treat effects (symptoms)
Langworthy	Sub-luxations	1902 (after September)	ASC announcement	"That at least 95% of all abnormal, deranged nerves are made so by sub-luxations of joints, more especially the spinal column"—quotes without attribution from D.D. Palmers's "Chiropractic Defined"
C.W. Burtch	Sub-luxations	October 1903	Backbone	Pressure is found to be caused almost invariably by sub-luxations of the spinal vertebrae

Box 4.2 Definitions of subluxation

... an historical concept ... not supported by any clinical research evidence that would allow claims to be made that it is the cause of disease or health concerns (General Chiropractic Council, 2010)

... a neurological imbalance or distortion in the body associated with adverse physiological responses and/or structural changes, which may become persistent or progressive. The most frequent site for the chiropractic correction of the subluxation is via the vertebral column. (Council on Chiropractic Practice, 2013)

... a self-perpetuating, central segmental motor control problem that involves a joint, such as a vertebral motion segment, that is not moving appropriately, resulting in ongoing maladaptive neural plastic changes that interfere with the central nervous system's ability to self-regulate, self-organize, adapt, repair and heal. (The Rubicon Group, 2017)

Box 4.3 Quotes about subluxation from a 1922 'Palmer catalog' (locations referred to descend from the upper neck to the lower back)

1. Slight subluxation at this point will cause so-called headaches, eye diseases, deafness, epilepsy, vertigo, insomnia wry neck, facial paralysis, locomotor ataxia, etc.
2. A slight subluxation of a vertebra in this part of the spine is the cause of so-called throat trouble, neuralgia, pain in the shoulders and arms, goitre, nervous prostration, grippe, dizziness, bleeding from nose, disorder of gums, catarrh, etc.
3. [Upper thoracic area] of the spine wherein subluxations will cause so-called bronchitis, felons, pain between the shoulder blades, rheumatism of the arms and shoulders, hay fever, writers' cramp, etc.
4. A vertebral subluxation at this point causes nervousness, heart disease, asthma, pneumonia, tuberculosis, difficult breathing, other lung troubles, etc.
5. Stomach and liver troubles, enlargement od spleen, pleurisy and a score of other troubles, so-called, are caused by subluxations of this part of the spine, sometimes so light as to remain unnoticed by other except the trained Chiropractor.
6. Here we find the cause of so-called gall stones, dyspepsia of upper bowels, fevers, shingles, hiccups, worms, etc.
7. Bright's disease, diabetes, floating kidney, skin disease, boils, erruptions and other diseases, so-called, are caused by nerves being pinched in the spinal openings at this point.
8. Regulations of such troubles as so-called appendicitis, peritonitis, lumbago, etc., follow Chiropractic adjustments at this point.

9. Why have so-called constipation, rectal troubles, sciatica, etc., when Chiropractic adjustments at this part of the spine will remove the cause?
10. A slight slippage of one or both innominate bones will likewise produce so-called sciatica, together with many "diseases" of the pelvis and lower extremities.

5

Spinal Manipulation

The principal functions of the spine: To support the head.
To support the ribs. To support the chiropractor.
—B. J. Palmer

The main reason why the myth of subluxation refuses to die is undoubtedly the fact that it provides the rationale for chiropractors' hallmark intervention: spinal manipulation and thus the financial basis for every chiropractor on the planet. The US 'Practice Analysis of Chiropractic 2020' showed that *manual chiropractic adjustment of the occiput, spine and/or pelvis was the treatment task performed with the highest frequency.*[1] Without these manipulations or adjustments, chiropractors would lose their identity and most of their income. Where there is a subluxation—and there is hardly a patient who has been found free of it—there is a reason to correct it with spinal manipulation and a reason to charge a fee.

Spinal manipulations have been used by bonesetters of several ancient cultures. Hippocrates used manipulative techniques and so did the ancient Egyptians. Before DD Palmer popularised chiropractic spinal manipulations, Andrew Still, the inventor of osteopathy, had developed his own techniques, and there is evidence that Palmer learnt from Still Chap. 8. As mentioned in Chap. 2, two of Palmer's students, Langworthy and Smith, accused DD

---

[1]Document available at https://mynbce.org/wp-content/uploads/2020/02/Executive-Summary-Practice-Analysis-of-Chiropractic-2020.pdf.

© Springer Nature Switzerland AG 2020
E. Ernst, *Chiropractic*,
https://doi.org/10.1007/978-3-030-53118-8_5

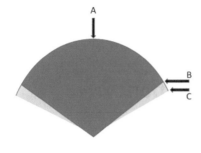

Schematic explanation of spinal manipulation

A joint can move from its neutral position (A) along one axis up to the limit within its physiological range of motion (B), the point to which a person can wilfully move it. With most techniques of spinal manipulation, the joint is passively forced slightly beyond this point (C) without causing injury by disrupting anatomical structures.

Fig. 5.1 The principle of spinal manipulation

Palmer of 'stealing' the technique of spinal manipulation from the Bohemian bonesetters of Iowa.

Today, many chiropractors still believe in DD Palmer's assumption that spinal manipulations work by adjusting subluxations, i.e. putting a bone that has gone out of place back into its correct position (Chap. 4). Others agree that this assumption is unrealistic and have thus postulated several other potential modes of action. So far, none of them offers a convincing explanation, and the clinical relevance of the observed effects remains uncertain (Box 5.1).

Despite this uncertainty, most chiropractors use spinal manipulations more regularly than any other treatment (Box 5.2) and patients consult chiropractors for a wide range of conditions (Box 5.3).

Palmer and his followers developed numerous methods of spinal manipulation (Fig. 5.1); some of the more widely used techniques include the following:

- The *diversified technique* involves a high velocity, low amplitude (HVLA) thrust usually delivered by hand. It is considered the most generic technique (see below). Its objective is to restore full movement and alignment of the spine and to normalise joint function.
- The *flexion distraction* involves the use of a specialized table that gently distracts or stretches the spine, and which allegedly allows the chiropractor to identify the area of abnormality while slightly flexing the spine in a pumping rhythm.
- The *activator method* involves the use of a handheld spring-loaded instrument which delivers a small impulse to the spine.

- The *atlas orthogonal technique* uses a percussion instrument on the upper spine to readjust subluxations. It is based on the teachings of BJ Palmer who advocated the 'Hole-In-One' version of spinal adjustment.
- The *Cox flexion-distraction* involves a manipulation aimed at adjusting vertebrae by applying a stretch to the lower spine, usually in a series of repetitive slow rocking motions.
- The *Gonstead adjustment* (or 'Palmer-Gonstead technique') is a HVLA adjustment employing an evaluation of the restricted joint and positioning of the patient's body. Specially designed chairs and tables are sometimes used to position the patient.
- The *Koren specific technique* was developed by Tedd Koren and involves the use of hands or an electric device, the "ArthroStim", for assessments and adjustments.
- *Release work* applies gentle pressure using the fingertips allegedly separating subluxated vertebrae with the aim of realigning them.
- The *sacro-occipital technique* involves placing wedges or blocks under the patient's pelvis, allowing gravity with the addition of low force manual pressure aimed at realigning the pelvis.
- The *Toggle drop* employs the chiropractor's crossed hands one on top of the other pressing quickly and firmly on a particular area of the spine while a section of the patient's bed, the drop table, gives way. The technique thus combines the chiropractor's force with that of gravity.

The above definitions of commonly used techniques are, of course, rather vague. In fact, the diversified technique is not a single method but a hotch-potch of techniques. This means that, even if one particular clinical trial tested the effectiveness of one specifically named chiropractic manipulation, such as the diversified technique, the chiropractors doing the treatments would most likely do what they believe is required for each individual patient. In other words, a plethora of different techniques are used as the chiropractor sees fit and no standardisation exists.

The range of techniques that any given chiropractor might use depends on their education, training and experience, the type of equipment available, the condition that is being treated, as well as other factors. There is no general agreement amongst chiropractors as to when to use which method of spinal manipulation. Similarly, there is little meaningful research into the effectiveness of different techniques. The few comparative studies that exist are usually of poor quality and their findings are thus less than reliable. The aim of one investigation, for instance, was to determine which techniques are used most frequently by Australian chiropractors to treat spinal musculoskeletal conditions.[2] Its findings confirm that the diversified technique was the first choice for most conditions. Flexion distraction was used almost as much as the diversified technique in the treatment of lumbar disc syndrome with radiculopathy and lumbar central stenosis. Seasoned chiropractors used more activator technique and soft tissue mobilisations and less diversified technique compared to their less experienced chiropractors.

When HVLA thrusts are applied, the joint is usually forced to move beyond its physiological limit of motion (Fig. 5.1). This often produces a noise that is clearly audible for the patient; this is often interpreted as the infallible sign that a vertebra was 'out of place' and has now been put back where it belongs. The notion is, however, incorrect. Belgian researchers conducted a survey aimed at examining the beliefs about this cracking sounds during HVLA thrusts in 100 individuals with and without personal experience of this technique.[3] The sounds were ascribed to vertebral repositioning by 49% of participants and to friction between two vertebras by 23% of participants. The sound was considered to indicate a successful spinal manipulation by 40% of participants. Yet, in reality, the audible pop has no therapeutic value. It is generated by a phenomenon called 'cavitation'. The HVLA thrust forces the joint to open up a little bit; this creates a vacuum inside the joint which, in turn, allows gas bubbles to form in the joint fluid.[4] And it is this formation of the gas bubbles that makes the popping sound.

Assessments of spinal manipulation by location are relevant not least because of their different risks. This report by the Canadian government provided an excellent assessment of the evidence for manipulations of the upper spine, the area where the potential for harm is greatest (Chap. 14).[5] Here are its key findings:

A total of 159 references were identified and cited in this review: 86 case reports/case series, 37 reviews of the literature, 9 randomized controlled trials, 6 surveys/qualitative studies, 5 case-control studies, 2 retrospective studies, 2 prospective studies and 12 others.

---

[2]Clijsters M, Fronzoni F, Jenkins H. Chiropractic treatment approaches for spinal musculoskeletal conditions: a cross-sectional survey. Chiropr Man Therap. 2014; 22(1):33. Published 2014 Oct 1. https://doi.org/10.1186/s12998-014-0033-8.

[3]Demoulin C, Baeri D, Toussaint G, Cagnie B, Beernaert A, Kaux J-F, Vanderthommen M.
   Beliefs in the population about cracking sounds produced during spinal manipulation. Joint Bone Spine, Volume 85, Issue 2, March 2018, pp. 239–242.

[4]Brodeur R. The audible release associated with joint manipulation. *J Manipulative Physiol Ther.* 1995; 18(3):155–164.

[5]Document available at https://www.gov.mb.ca/health/rhpa/docs/hnm.pdf.

Serious adverse events following cervical spinal manipulation (CSM) seem to be rare, whereas minor adverse events occur frequently.

Minor adverse events can include transient neurological symptoms, increased neck pain or stiffness, headache, tiredness and fatigue, dizziness or imbalance, extremity weakness, ringing in the ears, depression or anxiety, nausea or vomiting, blurred or impaired vision, and confusion or disorientation.

Serious adverse events following CSM can include the following: cerebrovascular injury such as cervical artery dissection, ischemic stroke, or transient ischemic attacks; neurological injury such as damage to nerves or spinal cord (including the dura mater); and musculoskeletal injury including injury to cervical vertebral discs (including herniation, protrusion, or prolapse), vertebrae fracture or subluxation (dislocation), spinal edema, or issues with the paravertebral muscles.

Rates of incidence of all serious adverse events following CSM range from 1 in 10,000 to 1 in several million cervical spine manipulations, but the literature generally agrees that serious adverse events are likely underreported.

The best available estimate of incidence of vertebral artery dissection of occlusion attributable to CSM is approximately 1.3 cases for every 100,000 persons <45 years of age receiving CSM within 1 week of manipulative therapy. The current best incidence estimate for vertebral dissection-caused stroke associated with CSM is 0.97 residents per 100,000.

While CSM is used by manual therapists for a large variety of indications including neck, upper back, and shoulder/arm pain, as well as headaches, the evidence seems to support CSM as a treatment of headache and neck pain only. However, whether CSM provides more benefit than spinal mobilization is still contentious.

A number of factors may make certain types of patients at higher risk for experiencing an adverse cerebrovascular event after CSM, including vertebral artery abnormalities or insufficiency, atherosclerotic or other vascular disease, hypertension, connective tissue disorders, receiving multiple manipulations in the last 4 weeks, receiving a first CSM treatment, visiting a primary care physician, and younger age. Patients who have experienced prior cervical trauma or neck pain may be at particularly higher risk of experiencing an adverse cerebrovascular event after CSM.

Conclusion: The current debate around CSM is notably polarized. Many authors stated that the risk of CSM does not outweigh the benefit, while others maintained that CSM is safe—especially in comparison to conventional treatments—and effective for treating certain conditions, particularly neck pain and headache. Because the current state of the literature may not yet be robust enough to inform definitive prohibitory or permissive policies around the application of CSM, an interim approach that balances both perspectives may involve the implementation of a harm-reduction strategy to mitigate potential harms of CSM until the evidence is more concrete. As noted by authors in the literature, approaches might include ensuring manual therapists are providing informed consent before treatment; that patients are provided with resources to aid in early recognition of a serious adverse event; and that regulatory bodies ensure the establishment of consistent definitions of adverse events for effective reporting and surveillance, institute rigorous protocol for identifying high-risk patients, and create detailed guidelines for appropriate application and contraindications of CSM. Most authors indicated that manipulation of the upper cervical spine should be reserved for carefully selected musculoskeletal conditions and that CSM should not be utilized in circumstances where there has not yet been sufficient evidence to establish benefit.

Spinal manipulations are employed not just by chiropractors, but also by several professions, including physiotherapists and osteopaths. For chiropractors, however, it is the hallmark therapy, and only chiropractors use spinal manipulation in the belief that they adjust vertebral subluxations which are the cause of many diseases (Chap. 4).

- Chiropractors use spinal manipulation for (almost) every patient.
- They use it for (almost) every condition.
- As they view the spine as one entity, they would correct subluxations independent of a patients site of pain; they might, for instance, manipulate the neck of a patient with back pain.
- They have developed most of the techniques (see above).
- Spinal manipulation is the focus of their theoretical education and practical training.
- All textbooks of chiropractic have a clear focus on spinal manipulation.
- Chiropractors conduct most of the research on spinal manipulation.
- Chiropractors are responsible for most of the adverse effects of spinal manipulation.

Most chiropractors claim that spinal manipulation must be specific and targeted at the affected (subluxated) joint. This notion is, however, debatable. A study from 2019 evaluated the effects of a targeted manipulative thrust versus a thrust applied generally to the lumbar region.[6] Sixty patients with low back pain were randomly allocated to two groups: one group received a targeted manipulative thrust (n = 29) and the other a general manipulation thrust (GT) (n = 31) to the lumbar spine. Thrust was either localised

---

[6]McCarthy CJ, Potter L, Oldham JA. Comparing targeted thrust manipulation with general thrust manipulation in patients with low back pain. A general approach is as effective as a specific one. A randomised controlled trial. *BMJ Open Sport & Exercise Medicine* 2019; **5**:e000514. https://doi.org/10.1136/bmjsem-2019-000514.

to a clinician-defined symptomatic spinal level or an equal force was applied through the whole lumbosacral region. The investigators measured pressure-pain thresholds (PPTs) using algometry and muscle activity (magnitude of stretch reflex) via surface electromyography. The results showed no between-group differences in self-reported pain or PPT. The authors concluded *that a GT procedure—applied without any specific targeting—was as effective in reducing participants' pain scores as targeted approaches.*

Similar findings originate from a study[7] demonstrating no significant difference in pain response to a general versus specific rotation, manipulation technique. The fact that 'targeted' manipulation is no better than 'general' manipulation challenges the need for current courses involving manual skill training and teaching of specific techniques. Assuming that GT was employed in the above studies as a placebo control, these studies suggest that the effects of spinal manipulation are largely or even entirely due to a placebo response (Chaps. 10 and 11).

Box 5.1 Some of the potential modes of action of spinal manipulation

- Spinal manipulation might affect the central nervous system by changing brain metabolism.[8]
- It might affect the inflow of sensory information to the central nervous system.[9]
- It might affect 'central facilitation', a phenomenon known to increase the receptive field of central neurons, enabling either subthreshold or innocuous stimuli access to central pain pathways.[7]
- I might alter central sensory processing by removing chemical stimuli from paraspinal tissues.[7]
- It might evoke paraspinal muscle reflexes and alters motoneuron excitability.[7]
- It might sympathetic nerve activity.[7]
- It might increase substance-p, neurotensin, oxytocin and interleukin levels.[10]

---

[7] Sutlive TG, Mabry LM, Easterling EJ, et al. Comparison of short-term response to two spinal manipulation techniques for patients with low back pain in a military beneficiary population. *Mil Med*. 2009; 174(7):750–756. https://doi.org/10.7205/milmed-d-02-4908.

[8] http://upflow.co/l/giQI/wordpress/2019/12/the-effect-of-spinal-manipulation-on-brain-neurometabolites-in-chronic-nonspecific-low-back-pain-patients.

[9] Pickar JG. Neurophysiological effects of spinal manipulation. *Spine J*. 2002; 2(5):357–371. https://doi.org/10.1016/s1529-9430(02)00400-x.

[10] Kovanur-Sampath K, Mani R, Cotter J, Gisselman AS, Tumilty S. Changes in biochemical markers following spinal manipulation-a systematic review and meta-analysis. *Musculoskelet Sci Pract*. 2017; 29:120–131. https://doi.org/10.1016/j.msksp.2017.04.004.

- It might increase moto-neurone excitability in the lower limb muscles.[11]
- It might decrease the discharge of muscle spindles.[12]
- It might exert inhibitory effects on temporal summation of pain.[13]

(The clinical relevance of any of these effects is uncertain).

Box 5.2 Frequency with which US chiropractors employ spinal manipulations in their daily practice[14]

Never 1.4%
1–6 times per year 0.4%
Once per month 1.0%
Once per week 1.1%
Once per day 2.0%
Many times per day 94%.

Box 5.3 Conditions for which US patients use spinal manipulations[15]

67% of all patients used it to treat a specific health condition
43% used it for general wellness or disease prevention
25% used it because it focuses on the whole person; their mind, body, and spirit
16% used it for improving their energy
11% used it for better immune function
5% used it to improve memory or concentration.

---

[11] Haavik H, Niazi IK, Jochumsen M, et al. Chiropractic spinal manipulation alters TMS induced I-wave excitability and shortens the cortical silent period. *J Electromyogr Kinesiol*. 2018; 42:24–35. https://doi.org/10.1016/j.jelekin.2018.06.010.

[12] Reed WR, Pickar JG, Sozio RS, Liebschner MAK, Little JW, Gudavalli MR. Characteristics of Paraspinal Muscle Spindle Response to Mechanically Assisted Spinal Manipulation: A Preliminary Report. *J Manipulative Physiol Ther*. 2017; 40(6):371–380. https://doi.org/10.1016/j.jmpt.2017.03.006.

[13] Randoll C, Gagnon-Normandin V, Tessier J, et al. The mechanism of back pain relief by spinal manipulation relies on decreased temporal summation of pain. *Neuroscience*. 2017; 349:220–228. https://doi.org/10.1016/j.neuroscience.2017.03.006.

[14] Document available at https://mynbce.org/wp-content/uploads/2020/02/Practice-Analysis-of-Chiropractic-2020-3.pdf.

[15] https://nccih.nih.gov/health/pain/spinemanipulation.htm.

6

Other Alternative Modalities Used by Chiropractors

Many chiropractic doctors offer natural adjunctive therapies, regimes and supplementation proven to effectively augment chiropractic treatment. Homeopathic medicine is included in this list.[1]

Chiropractors often stress that they use not just spinal manipulation but also a multitude of other modalities. And it's true, legally they are allowed to do more, much more. Obviously, the law differs from country to country but, generally speaking, the 'mixers' amongst the chiropractors (Chap. 2) employ many methods borrowed from physiotherapists, and practitioners of so-called alternative medicine (SCAM).

A 2014 survey[2] of US chiropractic practice laws revealed the highest number of services that could be performed in the following US states (the numbers in brackets signify the number of allowed services):

- Missouri (n = 92),
- New Mexico (n = 91),
- Kansas (n = 89),
- Utah (n = 89),
- Oklahoma (n = 88),
- Illinois (n = 87),
- Alabama (n = 86).

---

[1] https://www.dynamicchiropractic.com/mpacms/dc/article.php?id=8985.

[2] Chang M. The chiropractic scope of practice in the United States: a cross-sectional survey. *J Manipulative Physiol Ther.* 2014; 37(6):363–376. https://doi.org/10.1016/j.jmpt.2014.05.00.

© Springer Nature Switzerland AG 2020
E. Ernst, *Chiropractic,*
https://doi.org/10.1007/978-3-030-53118-8_6

And here are some of the allowed services that might be surprising:

- birth certificates
- death certificates
- premarital certificates
- recto-vaginal exam
- venepuncture
- intravenous injections
- prostatic exam
- genital exam
- hyperbaric chamber.

In the area of SCAM, most chiropractors employ an impressive array of so-called alternative medicines (SCAMs), some of which were originally even invented by chiropractors. In the following section, I will discuss some of the most used (SCAMs) and diagnostic techniques.

Acupuncture

Acupuncture is amongst the most popular SCAMs. Thirteen percent of US chiropractors[3] and about half of all Canadian chiropractors, for instance, seem to use acupuncture.[4] Yet, experts are still divided in their views about it. Some accept that acupuncture works for some conditions, while many others remain unconvinced.[5,6,7]

Some chiropractors employ the traditional Chinese approach, while others adhere to the principles of conventional medicine. Traditional Chinese acupuncture is based on the Taoist philosophy of the balance between two life-forces, 'yin and yang'. In contrast, the 'Western' approach is based on

---

[3]Document available at https://mynbce.org/wp-content/uploads/2020/02/Practice-Analysis-of-Chiropractic-2020-3.pdf.

[4]Carlesso LC, Macdermid JC, Gross AR, Walton DM, Santaguida PL. Treatment preferences amongst physical therapists and chiropractors for the management of neck pain: results of an international survey. *Chiropr Man Therap*. 2014; 22(1):11. Published 2014 Mar 24. https://doi.org/10.1186/2045-709X-22-11.

[5]Colquhoun D, Novella SP. Acupuncture is theatrical placebo. *Anesth Analg*. 2013; 116(6):1360–1363. https://doi.org/10.1213/ANE.0b013e31828f2d5e.

[6]He Y, Guo X, May BH, et al. Clinical Evidence for Association of Acupuncture and Acupressure With Improved Cancer Pain: A Systematic Review and Meta-Analysis [published online ahead of print, 2019 Dec 19]. *JAMA Oncol*. 2019; 6(2):271–278. https://doi.org/10.1001/jamaoncol.2019.5233.

[7]Paley CA, Johnson MI. Acupuncture for the Relief of Chronic Pain: A Synthesis of Systematic Reviews. *Medicina* **2020**, *56*, 6.

neurophysiological theories as to how acupuncture might work[8]; even though these may appear plausible, they are mere theories and constitute no proof for acupuncture's validity.

According to the traditional view, acupuncture is useful for virtually every condition affecting mankind. According to 'Western' acupuncturists, acupuncture is effective for a much smaller range of conditions, mostly chronic pain. Acupuncture is associated with a powerful placebo effect; it works better than a placebo only for very few (some say for no) conditions.[9] Most of the clinical trials of acupuncture originate from China, and several investigations have shown that almost all of them are positive, suggesting that the results of these studies must be questioned.

Mild to moderate side-effects of acupuncture occur in about 10% of all patients. Serious complications of acupuncture are on record as well: acupuncture needles can, for instance, injure vital organs like the lungs or the heart, and they can introduce infections into the body, e.g. hepatitis.[10] About 100 fatalities after acupuncture have been reported in the medical literature—a figure which, due to lack of a monitoring system, might disclose just the tip of an iceberg.

Given that, for most conditions, there is no good evidence that acupuncture works beyond a placebo effect, and that acupuncture can also cause harm, the risk of acupuncture usuallys fail to outweigh its benefits.

Applied Kinesiology

Applied kinesiology was developed by the US chiropractor, George J. Goodheart, Jr. (1918–2008). In a 2000 survey by the US National Board of Chiropractic Examiners, 43% of respondents stated that they used applied kinesiology, with similar numbers reported in Australia.[11] By manually assessing patients' relative muscular weaknesses, practitioners claim to diagnose their health problems and identify an effective therapy. The assumptions

---

[8]Zhao ZQ. Neural mechanism underlying acupuncture analgesia. *Prog Neurobiol*. 2008; 85(4):355–375. https://doi.org/10.1016/j.pneurobio.2008.05.004.

[9]Ernst E. Acupuncture: what does the most reliable evidence tell us? An update. *J Pain Symptom Manage*. 2012; 43(2):e11–e13. https://doi.org/10.1016/j.jpainsymman.2011.11.001.

[10]Ernst E, Lee MS, Choi TY. Acupuncture: does it alleviate pain and are there serious risks? A review of reviews. *Pain*. 2011; 152(4):755–764. https://doi.org/10.1016/j.pain.2010.11.004.

[11]Cuthbert SC, Goodheart GJ Jr. On the reliability and validity of manual muscle testing: a literature review. *Chiropr Osteopat*. 2007; 15:4. Published 2007 Mar 6. https://doi.org/10.1186/1746-1340-15-4.

upon which applied kinesiology are based are out of line with our understanding of how our bodies function.

Several scientific tests of applied kinesiology have been published, but most of them are of poor quality and therefore not reliable. A systematic review concluded that "there is insufficient evidence for diagnostic accuracy within kinesiology, the validity of muscle response and the effectiveness of kinesiology for any condition."[12] Even though applied kinesiology itself is unlikely to cause harm (other than to the patient's bank balance), it will lead to false-positive and false-negative diagnoses. This would obviously be harmful and, in extreme cases, it could even cost a patient's life.

Chelation Therapy

Chelation therapy is a well-established, potentially life-saving treatment for certain acute intoxications, e.g. by heavy metals. In chiropractic, it is used for all sorts of illnesses and for 'detox'. The principle of chelation therapy is to inject a chemical into the veins which binds ions that subsequently can be excreted. Some chiropractors promote chelation for a wide range of conditions ranging from arthritis to cardiovascular disease.

Several systematic reviews of the best evidence failed to show that this form of alternative chelation therapy is effective.[13] Chelation therapy can dramatically affect the electrolyte levels in the blood which carries serious risks. Several fatalities are on record. Chelation therapy is expensive; often the costs amount to several US$ 10 000 for one series of treatments. Considering the risks and the doubts about its effectiveness, the risk/benefit balance fails to be positive.

Cupping

Cupping is a form of SCAM with long traditions in several ancient cultures. There are two distinct forms: dry and wet cupping. Wet cupping involves scarring the skin with a sharp instrument and then applying a cup with a vacuum to suck blood from the wound. Dry cupping omits the scarring of

---

[12]Hall S, Lewith G, Brien S, Little P. A review of the literature in applied and specialised kinesiology. *Forsch Komplementmed*. 2008; 15(1):40–46. https://doi.org/10.1159/000112820.

[13]See, e.g. Sultan S, Murarka S, Jahangir A, Mookadam F, Tajik AJ, Jahangir A. Chelation therapy in cardiovascular disease: an update. *Expert Rev Clin Pharmacol*. 2017; 10(8):843–854. https://doi.org/10.1080/17512433.2017.1339601.

the skin and merely involves placing a vacuum cup over the skin. The suction then creates a subcutaneous haematoma which leaves the typical cupping mark that remains visible for several days. Dry cupping can be understood as a form of counter-irritation. While dry cupping is virtually risk-free, wet cupping can lead to serious infections and permanent scarring.

There has been a flurry of research into the effects of (mostly dry) cupping. Most clinical trials have been seriously flawed, and thus the conclusions of systematic reviews were inconclusive, e.g.:

- No explicit recommendation for or against the use of cupping for athletes can be made. More studies are necessary for conclusive judgment on the efficacy and safety of cupping in athletes.[14]
- Cupping therapy can significantly decrease the VAS scores and ODI scores for patients with LBP compared to the control management. High heterogeneity and risk of bias existing in studies limit the authenticity of the findings.[15]
- Only weak evidence can support the hypothesis that cupping therapy can effectively improve treatment efficacy and physical function in patients with knee osteoarthritis.[16]
- There are not enough trials to provide evidence for the effectiveness of cupping for stroke rehabilitation because most of the included trials compared the effects with unproven evidence and were not informative.[17]
- For wet cupping, there is far less research, and some of the results that emerged from clinical trials seem less than credible.[18]

On balance, the benefits of wet cupping do not seem to outweigh its risks, while for dry cupping the verdict remains uncertain.

---

[14]Bridgett R, Klose P, Duffield R, Mydock S, Lauche R. Effects of Cupping Therapy in Amateur and Professional Athletes: Systematic Review of Randomized Controlled Trials. *J Altern Complement Med*. 2018; 24(3):208–219. https://doi.org/10.1089/acm.2017.0191.

[15]Wang YT, Qi Y, Tang FY, et al. The effect of cupping therapy for low back pain: A meta-analysis based on existing randomized controlled trials. *J Back Musculoskelet Rehabil*. 2017; 30(6):1187–1195. https://doi.org/10.3233/BMR-169736.

[16]Li JQ, Guo W, Sun ZG, et al. Cupping therapy for treating knee osteoarthritis: The evidence from systematic review and meta-analysis. *Complement Ther Clin Pract*. 2017; 28:152–160. https://doi.org/10.1016/j.ctcp.2017.06.003.

[17]Lee MS, Choi TY, Shin BC, Han CH, Ernst E. Cupping for stroke rehabilitation: a systematic review. *J Neurol Sci*. 2010; 294(1–2):70–73. https://doi.org/10.1016/j.jns.2010.03.033.

[18]AlBedah A, Khalil M, Elolemy A, Hussein AA, AlQaed M, Al Mudaiheem A, Abutalib RA, Bazaid FM, Bafail AS, Essa A, and Bakrain MY. The Journal of Alternative and Complementary Medicine Vol. 21(8), 504–508, Aug 2015. https://doi.org/10.1089/acm.2015.0065.

Detox

Detox is short for 'detoxification'. In conventional medicine, the term is used for treatments that wean drug-dependent patients off their drugs. In SCAM, detox has become an umbrella term for a wide range of treatments that allegedly rid our bodies of toxins. Chiropractors only employ the latter form of detox.

Many chiropractors are wedded to the belief that our body is full of toxins which threaten our good health. Some go as far as claiming that *chiropractic adjustments actually release toxins that have built up in the body over time.*[19] Others employ all sorts of other SCAMs for the purpose. The toxins are said to originate from our body's own metabolism, from the environment, from prescribed drugs, or from the food we consume. Proponents of detox claim that their treatments help the body get rid of these toxins and thus restore health. In truth, they not only fail to be effective, but many are even harmful.

While the assumption that some toxins can accumulate in our body might be correct, the notion that any of the SCAM detox treatments effectively eliminate these toxins from our body lacks sound evidence. Our body has a powerful mechanism to achieve this aim on its own. One review of the literature concluded that the promotion of alternative detox treatments provides income for some entrepreneurs but has the potential to cause harm to patients and consumers.[20]

Dietary Supplements

For most chiropractors, the sale of supplements to patients provides a welcome stream of additional income (Box 6.1). Dietary supplements form a category of remedies that, in most countries, is poorly regulated and not allowed to make significant health claims. They can contain ingredients from minerals and vitamins, from plants and animal products. Therefore, generalisations across all types of supplements are impossible. The therapeutic claims that are being made for them (even though such claims are usually forbidden) range from weight loss to anti-aging, and from curing erectile dysfunction to preventing cardiovascular disease.[21]

---

[19]https://livelovefruit.com/chiropractic-adjustments-good-for-detox-positive-energy-and-health/.

[20]Ernst E. Alternative detox. *Br Med Bull*. 2012; 101:33–38. https://doi.org/10.1093/bmb/lds002.

[21]Cohen PA, Ernst E. Safety of herbal supplements: a guide for cardiologists. *Cardiovasc Ther*. 2010; 28(4):246–253. https://doi.org/10.1111/j.1755-5922.2010.00193.x.

Although they are usually promoted as natural and safe, dietary supplements do not necessarily have either of these qualities. Most are not supported by sound evidence of efficacy. Numerous investigations have shown that some supplements fail to contain the amount of the ingredient claimed by the manufacturer. Similarly, supplements have been shown repeatedly to be contaminated with toxic materials or adulterated with synthetic drugs.[22]

Energy Healing

Energy healing is an umbrella term for a range of paranormal healing practices that include spiritual healing Reiki, Therapeutic Tough and many more. Arguably, chiropractic, as viewed by its inventor, DD Palmer, is also a form of energy healing (Chap. 2). The common denominator for all forms of energy healing is the belief in a mystical 'energy' (Palmer's 'innate') that can be used for therapeutic purposes. Energy healing has existed in many ancient cultures, and the 'New Age' movement has brought about its revival. Today energy healing systems are part of many chiropractors' practice.

The 'energy' that healers refer to is not measurable and lacks biological plausibility. Yet, energy healing has attracted a surprisingly high level of research activity. The methodologically best trials of energy healing fail to demonstrate that it generates effects beyond placebo.[23] Even though energy healing is per se harmless, it can do untold damage, not least because it significantly undermines rational thought in our societies.

Homeopathy

Homeopathy was invented by the German physician, Samuel Hahnemann (1755–1843) (Fig. 6.2). At the time, conventional treatments of this period were often more dangerous than the disease they were supposed to cure. Consequently, homeopathy was repeatedly shown to be superior to 'allopathy' (a term coined by Hahnemann to denigrate conventional medicine) and

---

[22]Posadzki P, Watson L, Ernst E. Contamination and adulteration of herbal medicinal products (HMPs): an overview of systematic reviews. *Eur J Clin Pharmacol*. 2013; 69(3):295–307. https://doi.org/10.1007/s00228-012-1353-z.

[23]Ernst E. Distant healing—an "update" of a systematic review. *Wien Klin Wochenschr*. 2003; 115(7–8):241–245. https://doi.org/10.1007/BF03040322.

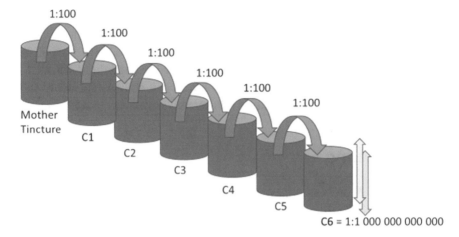

Fig. 6.1 Schematic explanation of the process of 'potentisation'. The original material (mother tincture) is diluted at a ratio of 1:100 to generate a C1 potency which is diluted to give a C2 potency, etc., etc. At each dilution, the remedy is vigorously shaken (succussed), as indicated by the yellow arrows. Homeopaths believe that the process of potentisation renders a remedy not less but more potent

Hahnemann's treatments were an almost instant, worldwide success.[24] Some chiropractors, particularly those from the US, employ homeopathy.

Many consumers confuse homeopathy with herbal medicine; yet the two are fundamentally different. Herbal medicines are plant extracts with potentially active ingredients. Homeopathic remedies can be based on plants or any other material and are typically so dilute that they contain not a single molecule of the substance advertised on the bottle (Fig. 6.1). The most frequently used dilution (homeopaths call them 'potencies') is a 'C30'; a C30-potency has been diluted 30 times at a ratio of 1:100. This means that one drop of the starting material is dissolved in 1 000 000 000 000 000 000 000 000 000 000 000 000 000 000 000 000 000 000 000 000 drops of diluent (usually a water/alcohol mixture)—and that equates to less than one molecule of the original substance for all the molecules in the universe Fig. 6.1.

Homeopaths claim that their remedies work via some 'energy' or 'vital force' and that the process of preparing the homeopathic dilutions (it involves vigorously shaking the mixtures at each dilution step) transfers this 'energy' or information from one to the next dilution. They also believe that the process of diluting and agitating their remedies, which they call potentisation, renders them not less but more potent. Homeopathic remedies are usually prescribed

---

[24] Ernst E. Homeopathy. The Undiluted Facts. Including a Comprehensive A–Z Lexicon, Springer (2016).

according to the 'like cures like' principle: if, for instance, a patient suffers from runny eyes, a homeopath might prescribe a remedy made of onion, because onion makes our eyes water. The assumptions of homeopathy contradict the known laws of nature. In other words, we do not fail to comprehend how homeopathy works, but we understand that it cannot work unless the known laws of nature are wrong.

Today, some 500 clinical trials of homeopathy have been published. The totality of this evidence fails to show that homeopathic remedies are more than placebos.[25] Yet, many patients undeniably do get better after taking homeopathic remedies. The best evidence available today shows that this improvement is unrelated to the homeopathic remedy per se, but the result of a lengthy, empathetic, compassionate encounter with a homeopath, a placebo-response or other factors which experts often call 'context effects'.[26] Whenever homeopaths advise their patients (as they often do) to forgo effective conventional treatments, they are likely to do harm. This phenomenon is best documented in relation to the advice of many homeopaths against immunisations.[27]

Massage

While most alternative types of massage are based on concepts that are out of line with our modern medical knowledge, Swedish massage is based on anatomy, physiology, etc., and its mode of action is arguably more plausible than that of all other types of massage therapy (Fig. 6.2).

It is often performed by a massage therapist, osteopath or chiropractor who would employ various techniques: effleurage (long smooth strokes), petrissage (kneading, rolling, and lifting), friction (wringing or small circular movements), tapotement (percussion), and vibration (rocking and shaking movements). In most European countries, massage is part of conventional healthcare, while elsewhere it is usually viewed as an alternative therapy.[28]

[25] Ernst E. Homeopathy: what does the "best" evidence tell us? *Med J Aust.* 2010; 192(8):458–460.

[26] Brien S, Lachance L, Prescott P, McDermott C, Lewith G. Homeopathy has clinical benefits in rheumatoid arthritis patients that are attributable to the consultation process but not the homeopathic remedy: a randomized controlled clinical trial. *Rheumatology (Oxford).* 2011; 50(6):1070–1082. https://doi.org/10.1093/rheumatology/keq234.

[27] Schmidt K, Ernst E. MMR vaccination advice over the Internet. *Vaccine.* 2003; 21(11–12):1044–1047. https://doi.org/10.1016/s0264-410x(02)00628-x.

[28] Posadzki P, Watson LK, Alotaibi A, Ernst E. Prevalence of use of complementary and alternative medicine (CAM) by patients/consumers in the UK: systematic review of surveys. *Clin Med (Lond).* 2013; 13(2):126–131. https://doi.org/10.7861/clinmedicine.13-2-126.

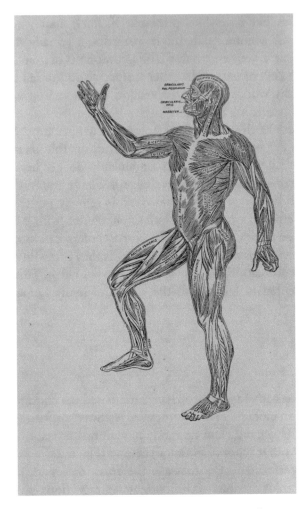

Fig. 6.2 Illustration of the human musculoskeletal system from an early massage handbook. *Source* US National Library of Medicine

Massage is advocated mainly to increase well-being, for relaxation and for musculoskeletal pain. Several systematic reviews of clinical trials have become available. However, their findings must be interpreted with caution, since the types of massage therapy used are often not clearly specified and differentiated. Their results vary according to the specific indication.

- One overview identified 31 systematic reviews of massage for pain control, of which 21 were considered high-quality. Findings from high-quality systematic reviews describe potential benefits of massage for pain indications including labour, shoulder, neck, back, cancer, fibromyalgia, and temporomandibular disorder.[29]
- A systematic review showed that massage therapy has promise for caner palliation: massage can alleviate a wide range of symptoms: pain, nausea, anxiety, depression, anger, stress and fatigue. However, the methodological quality of the included studies was poor, a fact that prevented definitive conclusions.[30]
- Another systematic review concluded that there is currently a lack of evidence to support the assertion that massage is effective for depression.[31]

Adverse effects of massage are mild and infrequent; a systematic review concluded that massage is not entirely risk free. However, serious adverse events are probably rare.[32]

Mind-Body Therapies

Mind-body therapies include a range of treatments that claim to exert a positive influence on health via the mind. Most mind-body therapies are supposed to induce a degree of relaxation which, in turn, can lead to a reduction of symptoms. Mind-body therapies deal with common experiences that cause distress around cancer diagnosis, treatment, and survivorship including loss of control, uncertainty about the future, fears of recurrence, and a range of physical and psychological symptoms including depression, anxiety, insomnia, and fatigue.[33] They are usually not curative but symptomatic treat-

---

[29] Miake-Lye I, Lee J, Lugar T, et al. *Massage for Pain: An Evidence Map.* Washington (DC): Department of Veterans Affairs (US); 2016.

[30] Ernst E. Massage therapy for cancer palliation and supportive care: a systematic review of randomised clinical trials. *Support Care Cancer.* 2009; 17(4):333–337. https://doi.org/10.1007/s00 520-008-0569-z.

[31] Coelho HF, Boddy K, Ernst E. Massage therapy for the treatment of depression: a systematic review. *Int J Clin Pract.* 2008; 62(2):325–333. https://doi.org/10.1111/j.1742-1241.2007.01553.x.

[32] Ernst E. The safety of massage therapy. *Rheumatology (Oxford).* 2003; 42(9):1101–1106. https://doi.org/10.1093/rheumatology/keg306.

ments and are rarely used as sole therapies. For several mind-body therapies, the evidence is encouraging but rarely compelling. This is usually due to

- a paucity of studies,
- the methodological problems encountered when conducting such clinical trials,
- the low quality of the existing trials,
- the lack of research funds in this area.

Vega Test

The Vega test (or electrodermal testing) involves an electronic device used by some chiropractors for diagnosing diseases. The Vega test is a development based on the electroacupuncture according to Voll. It was originally developed as an aid in prescribing homeopathic remedies. Today, the Vega test is popular with some chiropractors, predominantly for diagnosing allergies.

The Vega test is based on the assumption that changes in electrical impedance of the skin occur at the site of an acupuncture point in response to substances placed in an electrical circuit. The suggested mechanism of the Vega test has been summarised as "quantum biology." The only rigorous test of the Vega machine available to date concluded that it cannot be used to diagnose environmental allergies.[34] Several medical associations have advised against using the Vega test, including the National Institute for Health and Clinical Excellence, the Australian College of Allergy, the Australasian Society of Clinical Immunology and Allergy, the American Academy of Allergy, Asthma and Immunology and the Allergy Society of South Africa. False-positive and false-negative diagnoses are likely and have the potential to cause serious harm.

---

[33]Carlson LE. Distress Management Through Mind-Body Therapies in Oncology. *J Natl Cancer Inst Monogr*. 2017; 2017(52). https://doi.org/10.1093/jncimonographs/lgx009.

[34]Lewith GT, Kenyon JN, Broomfield J, Prescott P, Goddard J, Holgate ST. Is electrodermal testing as effective as skin prick tests for diagnosing allergies? A double blind, randomised block design study. *BMJ*. 2001; 322(7279):131–134. https://doi.org/10.1136/bmj.322.7279.131.

Box 6.1 Frequency with which US chiropractors use dietary supplements in their daily practice[35]

Never 29.7%
1–6 times per year 7.7%
Once per month 10.8%
Once per week 15.2%
Once per day 14.8%
Many times per day 21.8%.

---

[35] Document available at https://mynbce.org/wp-content/uploads/2020/02/Practice-Analysis-of-Chiropractic-2020-3.pdf.

7

Therapeutic Claims by (and Ambitions of) Chiropractors

[Chiropractic] will lessen disease, poverty and crime, empty our jails, penitentiaries and insane asylums and assist us to prepare for the existence beyond the transition called death.
—DD Palmer

Most consumers assume that chiropractors treat sore backs and necks but little else. Yet, this notion is far from correct. We have already seen that chiropractic started out as a profession believing in treating virtually every condition (Chap. 2), and this notion is still deeply engrained in the minds of many chiropractors. Box 7.1 lists a few conditions which are not readily associated with chiropractic but which, according to a recent survey, are nevertheless regularly treated by US chiropractors.[1] In this chapter I will discuss some of the claims made by chiropractors. In Chaps. 10–13, we will discuss details about the scientific evidence related to such claims.

To provide a flavour of the grossly over-optimistic attitude that is oddly typical of chiropractors, here is an excerpt from a press-release about a book entitled "Beyond the Back: The Chiropractic Alternative For Conditions Beyond Back Pain"[2]

---

[1] Document available at https://mynbce.org/wp-content/uploads/2020/02/Practice-Analysis-of-Chiropractic-2020-3.pdf.

[2] http://www.digitaljournal.com/pr/3183528.

© Springer Nature Switzerland AG 2020
E. Ernst, *Chiropractic*,
https://doi.org/10.1007/978-3-030-53118-8_7

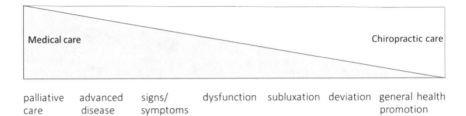

| palliative care | advanced disease | signs/ symptoms | dysfunction | subluxation | deviation | general health promotion |

Fig. 7.1 The scope of practice as seen by the US National Board of Chiropractic examiners (figure modified after the 'Practice Analysis of Chiropractic 2020' (see footnote 1))

> Beyond the Back focuses on how Chiropractic care can do so much more than just alleviate back pain. From avoiding knee surgery to resolving athletic injuries, chiropractic care is a 100% natural health solution for a wide variety of conditions… In fact, in some cases, chiropractors can help their patients get off medications entirely and even avoid surgery.

Chiropractors currently like to consider themselves as primary care physicians (Fig. 7.1). Already in 2002, a paper concluded that *the primary care chiropractic physician is a viable and important part of the primary health care delivery system, with many chiropractic physicians currently prepared to participate effectively and competently in primary care.*[3] In recent years, the American Chiropractic Association (ACA) have ramped up their lobbying on having chiropractors recognised as primary care physicians:

> The American Chiropractic Association (ACA) House of Delegates (HOD) met … and approved a resolution to make Medicare parity a top legislative and regulatory priority for the Association. The resolution, adopted by the House members, emphasizes the importance of allowing chiropractic physicians to practice and be reimbursed for the fullest extent of their licensure, training and competencies.[4]

And this ambition is not confined to the US. The 'Alliance of UK Chiropractors' stated in 2020:

> The best model for the provision of Chiropractic service to NHS patients would be one of independent assessment and self- referral, much like that

---

[3]Kremer RG, Duenas R, McGuckin B. Defining primary care and the chiropractic physicians' role in the evolving health care system. *J Chiropr Med*. 2002; 1(1):3–8. http://doi.org/10.1016/S0899-3467(07)60021-4.

[4]https://www.acatoday.org/News-Publications/Newsroom/News-Releases/ACA-Delegates-Declare-Medicare-Parity-a-Top-Priority.

which currently exists between a patient and their registered GP. Using this model the patient can either self-refer for assessment by their Chiropractor or their GP can refer them for care. This would result in quicker resolutions and better clinical outcomes for the patient and a reduced cost of care (i.e. better value) for the NHS. The Chiropractor would therefore be working in co-operation with the NHS and the GP in particular.[5]

But the wishful thinking does not stop there. Recently, the ACA published a document entitled 'Guidelines for Disaster Service by Doctors of Chiropractic'. Here are a few short quotes from this document:

> … Doctors of Chiropractic are uniquely qualified to serve in emergency situations in various capacities … their assessment and treatments can be performed in austere environments, on site or at staging areas providing rapid attention to the injury, accelerating healing and often decreasing or substituting the need for pharmaceutical intervention…Through their education as primary care physicians, Doctors of Chiropractic have demonstrated competence in first aid and resuscitation skills and are able to assess, diagnose and triage so they may serve as first responders in the immediate care of victims at a disaster site…During and after the disaster, the local Doctors of Chiropractic should interface with the state association and ACA to report on execution of action and outcome of the situation, make suggestions for response to future disasters and report any significant contacts made.

In the interest of truth, it is necessary to provide a few corrections and clarifications:

- Chiropractors are not medical doctors; to use the title in a medical context is misleading, to use it in the context of medical emergencies seems reckless.
- Chiropractors are not qualified to serve in emergency situations.
- There is no good evidence that chiropractic can accelerate healing of any medical condition.
- No robust evidence exists that chiropractic decreases the need for pharmaceutical interventions in emergency situations.
- Chiropractors are not trained to diagnose the complex or life-threatening conditions that occur in disaster situations.
- Chiropractors are not qualified or trained to report on execution of action and outcome of disaster situation.

---

[5]Document available at https://cdn.vortala.com/static/uploads/3/aukc-chiropractic-brochure_compressed.pdf.

- Chiropractors are not competent to make suggestions for response to future disasters.

Like DD Palmer, their founding father, many chiropractors have an aversion to drugs (Chap. 1). The distrust of the influence of 'Big Pharma' is an important motivator for consumers to try SCAM. Data from Spain, for instance, confirm that it has a significant effect on viewing SCAM as more effective.[6] Chiropractors' distrust in pharmacological treatments is all the more surprising considering that some chiropractors have recently started lobbying to get prescription rights. A recent article explains[7]:

> A legislative proposal that would allow Wisconsin chiropractors to prescribe narcotics has divided those in the profession and pitted those of them who support the idea against medical doctors. At a hearing on the bill Tuesday, representatives form the Wisconsin Chiropractic Association said back pain is a common reason people go see a medical doctor, but they argue that chiropractors with additional training could be helping those patients instead. Under the bill, chiropractors would be able to write prescriptions for painkillers and administer anesthesia under the direction of a physician.

> … some practicing chiropractors like Jason Mackey, with Leutke Storm Mackey Chiropractic in Madison, argue that medical fields evolve: "We have always had change throughout the course of our professsion." Mackey said there has been pushback with previous changes, like using X-ray or certain therapies and recommending vitamins.

An article by Canadian and Swiss chiropractors attempted to sum up the arguments for and against this prescription rights for chiropractors in more detail[8]:

Arguments in favour of prescription rights for chiropractors

- Such privileges would be in line with current evidence-based practice. Currently, most international guidelines recommend, alongside prescription medication, a course of manual therapy and/or exercise as well as

---

[6]Lobera J, Rogero-García J. Scientific Appearance and Homeopathy. Determinants of Trust in Complementary and Alternative Medicine [published online ahead of print, 2020 Apr 14]. *Health Commun.* 2020;1-8. https://doi.org/10.1080/10410236.2020.1750764.

[7]https://www.circleofdocs.com/some-chiropractors-push-for-ability-to-prescribe-painkillers-what-say-you/.

[8]Emary PC, Houweling TA, Wangler M, Burnie SJ, Hood KJ, Erwin WM. A commentary on the implications of medication prescription rights for the chiropractic profession. *Chiropr Man Therap.* 2016;24(1):33. Published 2016 Aug 24. http://doi.org/10.1186/s12998-016-0114-y.

education and reassurance as part of a multi-modal approach to managing various spine-related and other MSK conditions.

- Limited medication prescription privileges would be consistent with chiropractors' general experience and practice behaviour. Many clinicians tend to recommend OTC medications to their patients in practice.
- A more comprehensive treatment approach offered by chiropractors could potentially lead to a reduction in healthcare costs by providing additional specialized health care options for the treatment of MSK conditions. Namely, if patients consult one central practitioner who can effectively address and provide a range of treatment modalities for MSK pain-related matters, the number of visits to providers might be reduced, thereby resulting in better resource allocation.
- Limited medication prescription rights could lead to improved cultural authority for chiropractors and better integration within the healthcare system.
- With these privileges, chiropractors could have a positive influence on public health. For instance, analgesics and NSAIDs are widely used and potentially misused by the general public, and users are often unaware of the potential side effects that such medication may cause.

Arguments against prescription rights for chiropractors

- Chiropractors and their governing bodies would start reaching out to politicians and third-party payers to promote the benefits of making such changes to the existing healthcare system.
- Additional research may be needed to better understand the consequences of such changes and provide leverage for discussions with healthcare stakeholders.
- Existing healthcare legislation needs to be amended in order to regulate medication prescription by chiropractors.
- There is a need to focus on the curriculum of chiropractors. Inadequate knowledge and competence can result in harm to patients; therefore, appropriate and robust continuing education and training would be an absolute requirement.
- Another important issue to consider relates to the divisiveness around this topic within the profession. In fact, some have argued that the right to prescribe medication in chiropractic practice is the profession's most divisive issue. Some have argued that further incorporation of prescription rights into the chiropractic scope of practice will negatively impact the distinct professional brand and identity of chiropractic.

- Such privileges would increase chiropractors' professional responsibilities. For example, if given limited prescriptive authority, chiropractors would be required to recognize and monitor medication side effects in their patients.
- Prior to medication prescription rights being incorporated into the chiropractic scope of practice worldwide, further discussions need to take place around the breadth of such privileges for the chiropractic profession.

Some of these arguments are clearly spurious, particularly those in favour of prescription rights. Moreover, they are incomplete; here are a few additional thoughts:

- Patients might be put at risk by chiropractors who are not trained in prescribing medicines.
- More unnecessary NAISDs would be prescribed which obviously could harm some patients.
- The vast majority of the drugs in question is already available without a prescription.
- Healthcare costs would increase (just as plausible as the opposite argument made above).
- Prescribing rights would give legitimacy to a profession that arguably does not deserve it (Chaps. 10–15).
- Chiropractors would continue their lobbying and soon demand the prescription rights to be extended to other classes of drugs.

The UK Advertising Standards Authority (ASA) recently published a statement[9] outlining which claims UK chiropractors are allowed to make and which are likely to get them into conflict with the ASA. Here are a few excerpts:

> In 2017 the ASA carried out an evidence review on the use of multi-modal approaches used in Chiropractic in treating sciatica, whiplash and sports injuries as well as the treatment of babies, children and pregnant women as specific patient groups. The subsequent ASA Guidance explains in more detail the types of claims (including phraseology) that are likely to be acceptable for chiropractors to make in their advertising and those which are not. We recommend chiropractors consider this CAP advice and the ASA Guidance together when making treatment claims in advertising.

---

[9]https://www.asa.org.uk/advice-online/health-chiropractic.html.

Based on all evidence submitted and reviewed to date, the ASA and CAP accept that chiropractors may claim to treat the following conditions:

- *Ankle sprain (short term management)*
- *Cramp*
- *Elbow pain and tennis elbow (lateral epicondylitis) arising from associated musculoskeletal conditions of the back and neck, but not isolated occurrences*
- *Headache arising from the neck (cervicogenic)*
- *Inability to relax*
- *Joint pains*
- *Joint pains including hip and knee pain from osteoarthritis as an adjunct to core OA treatments and exercise*
- *General, acute & chronic backache, back pain (not arising from injury or accident)*
- *Generalised aches and pains*
- *Lumbago*
- *Mechanical neck pain (as opposed to neck pain following injury i.e. whiplash)*
- *Migraine prevention*
- *Minor sports injuries and tensions*
- *Muscle spasms*
- *Plantar fasciitis (short term management)*
- *Rotator cuff injuries, disease or disorders*
- *Sciatica*
- *Shoulder complaints (dysfunction, disorders and pain)*
- *Soft tissue disorders of the shoulder.*

For most of these conditions, there is no good evidence at all (Chaps. 11 and 12), and the list begs the question, which conditions do chiropractors actually treat? A 2016 survey of Australian chiropractors[10] indicated that chiropractors see patients for (the percentage figures refer to the percentages of patients):

- maintenance: 39%
- spinal problems: 33%
- neck problems: 18%
- shoulder problems: 6%
- headache: 6%
- hip problems: 3%

---

[10]Charity MJ, Britt HC, Walker BF, et al. Who consults chiropractors in Victoria, Australia?: Reasons for attending, general health and lifestyle habits of chiropractic patients. *Chiropr Man Therap.* 2016;24(1):28. Published 2016 Sep 1. https://doi.org/10.1186/s12998-016-0110-2.

- leg problems: 3%
- muscle problems: 3%
- knee problems: 2%.

Remarkably, for none of these conditions is there enough good evidence to justify this decision (Chaps. 10–12).

Many chiropractors claim that their spinal manipulations cause a reduction not just of back pain but of pain perception in general. They often call this phenomenon 'manipulation-induced hypoalgesia' (MIH). It is unknown, however, whether MIH is a real and clinically relevant treatment effect or mere wishful thinking aimed at increasing the cash-flow of chiropractors. A recent systematic review was aimed at finding out;[11] the authors investigated changes in sensory measures following high-velocity low-amplitude spinal manipulation in musculoskeletal pain populations. Fifteen studies were included. Thirteen measured pressure pain thresholds. Change in pressure pain threshold after spinal manipulation compared to sham revealed no significant difference. The authors concluded that *there was low quality evidence of no significant difference compared to sham manipulation.* In other words, there is no robust evidence that MIH is real.

However, many chiropractors disagree and claim that they have the answer to the growing problem of opioid overuse. For instance, Alison Dantas, CEO of the Canadian Chiropractic Association (CCA), stated:

Chiropractic services are an important alternative to opioid prescribing… We are committed to working collaboratively to develop referral tools and guidelines for prescribing professions that can help to prioritize non-pharmacological approaches for pain management and reduce the pressure to prescribe… We are looking to build an understanding of how to better integrate care that is already available in communities across Canada… Integrating chiropractors into interprofessional care teams has been shown to reduce the use of pharmacotherapies and improve overall health outcomes. This effort is even more important now because the new draft Canadian prescribing guidelines strongly discourage first use of opioids.[12]

---

[11] Charity MJ, Britt HC, Walker BF, et al. Who consults chiropractors in Victoria, Australia?: Reasons for attending, general health and lifestyle habits of chiropractic patients. *Chiropr Man Therap.* 2016;24(1):28. Published 2016 Sep 1. http://doi.org/10.1186/s12998-016-0110-2.

[12] https://www.newswire.ca/news-releases/the-canadian-chiropractic-association-applauds-health-can adas-action-to-address-the-opioid-crisis-612194283.html.

This, of course, begs several questions:

- Do chiropractors not know that there is considerable doubt over the efficacy of chiropractic manipulation for back pain (Chap. 11)?
- Do they not know that, for all other indications, the evidence is even less convincing (Chap. 12)?
- What evidence exists to assume they are in a position to 'develop referral tools and guidelines for prescribing professions'?
- Do they forget that they have never had prescribing rights, understand little about pharmacology, and have traditionally been against drugs of all kinds?
- Do they really believe there is good evidence showing that 'integrating chiropractors into interprofessional care teams… reduce(s) the use of pharmacotherapies and improve overall health outcomes'?

Many chiropractors tell their patients that vaccinations are not necessary, as long as they receive regular spinal adjustments (Chap. 15). This claim is based on the assumption that spinal manipulations stimulate the immune system. The text published on this website[13] is but one example of many:

> The nervous system and immune system are hardwired and work together to create optimal responses for the body to adapt and heal appropriately. Neural dysfunctions due to spinal misalignments are stressful to the body and cause abnormal changes that lead to a poorly coordinated immune response. Chiropractic adjustments have been shown to boost the coordinated responses of the nervous system and immune system…
>
> Subluxation is the term for misalignments of the spine that cause compression and irritation of nerve pathways affecting organ systems of the body. Subluxations are an example of physical nerve stress that affects neuronal control. According to researchers, such stressful conditions lead to altered measures of immune function & increased susceptibility to a variety of diseases.
>
> Inflammatory based disease is influenced by both the nervous, endocrine, and immune systems. Nerve stimulation directly affects the growth and function of inflammatory cells. Researchers found that dysfunction in this pathway results in the development of various inflammatory syndromes such as rheumatoid arthritis and behavioral syndromes such as depression. Additionally, this dysfunctional neuro-endo-immune response plays a significant role in immune-compromised conditions such as chronic infections and cancer.
>
> Wellness based chiropractors analyze the spine for subluxations and give corrective adjustments to reduce the stress on the nervous system. A 1992

---

[13] https://www.naturalnews.com/031206_chiropractic_immunity.html.

research group found that when a thoracic adjustment was applied to a subluxated area the white blood cell (neutrophil) count collected rose significantly.

And other websites go even further[14]:

The best way to prevent meningitis, and other illness, is to develop a robust immune system. The most important element in developing a robust immune system is optimum communication between all systems of the body. Chiropractic does this. The goal of chiropractic is to remove interference in the nervous system, the system that controls and coordinates all other parts of the body. Interference is caused by subluxations or misalignments in the spine. When subluxations are corrected, the body's nervous system functions optimally and boosts the immune functioning. In fact, individuals who receive chiropractic care have 200% greater immune competence than individuals who don't. This is why it is vital to receive regular chiropractic adjustments…

Looking at the actual research that might support such claims, we find that that it is scarce, flimsy and unconvincing. Nobody has yet shown that people who receive regular chiropractic adjustments are protected from infections. Unless such an effect can be demonstrated beyond reasonable doubt, we should be highly sceptical of the claim that chiropractic care stimulates the immune system. But some chiropractors disagree; this article,[15] for instance, claims that, in 1918, chiropractic proved itself to be the method of choice for treating the flu! Here is a short quote from it:

Chiropractors got fantastic results from influenza patients while those under medical care died like flies all around. Statistics reflect a most amazing, almost miraculous state of affairs. The medical profession was practically helpless with the flu victims but chiropractors seemed able to do no wrong."

In Davenport, Iowa, 50 medical doctors treated 4,953 cases, with 274 deaths. In the same city, 150 chiropractors including students and faculty of the Palmer School of Chiropractic, treated 1,635 cases with only one death.

"In the state of Iowa, medical doctors treated 93,590 patients, with 6,116 deaths – a loss of one patient out of every 15. In the same state, excluding Davenport, 4,735 patients were treated by chiropractors with a loss of only 6 cases – a loss of one patient out of every 789.

---

[14]https://edzardernst.com/2015/07/do-regular-chiropractic-adjustments-stimulate-the-immune-system-or-just-the-chiropractors-cash-flow/ (the original site has been removed but is discussed here).
[15]https://www.circleofdocs.com/1918-influenza-epidemic-and-chiropractic-care/.

National figures show that 1,142 chiropractors treated 46,394 patients for influenza during 1918, with a loss of 54 patients – one out of every 886.

Reports show that in New York City, during the influenza epidemic of 1918, out of every 10,000 cases medically treated, 950 died; and in every 10,000 pneumonia cases medically treated 6,400 died. These figures are exact, for in that city these are reportable diseases.

In the same epidemic, under drugless methods, only 25 patients died of influenza out of every 10,000 cases; and only 100 patients died of pneumonia out of every 10,000 cases…

In the same epidemic reports show that chiropractors in Oklahoma treated 3,490 cases of influenza with only 7 deaths. But the best part of this is, in Oklahoma there is a clear record showing that chiropractors were called in 233 cases where medical doctors had cared for the patients, and finally gave them up as lost. The chiropractors saved all these lost cases but 25.

Chiropractors who believe such claims seem to be unable to differentiate pseudoscience from science, anecdote from evidence, and cause from effect. In the typical epidemiological case/control study, one large group of patients [A] is retrospectively compared to another group [B]. In our case, group A was treated by chiropractors, while group B received the treatments available at the time. It is true that several of such reports seemed to suggest that chiropractic works. But this does not prove much; the result might have been due to a range of circumstances, for instance:

- group A might have been less ill than group B,
- group A might have been richer and therefore better nourished,
- group A might have benefitted from better hygiene,
- group A might have received better care, e.g. hydration,
- group B might have received treatments that made the situation not better but worse.

Because of the retrospective nature of these studies, there is no way to account for these and many other factors confounding the outcome. This means that epidemiological studies of this type can generate interesting results which, in turn, need testing in properly controlled studies where these confounding factors are adequately controlled for. Without such tests, they are next to worthless.

Some chiropractors also claim to be able to treat patients suffering from weight problems. But is this claim any more than a marketing gimmick? A team of chiropractors performed a retrospective file analysis of patient files attending their 13-week weight loss program.[16] The program consisted of chiropractic adjustments/spinal manipulative therapy augmented with diet/nutritional intervention, exercise and one-on-one counselling. Sixteen of 30 people enrolled completed the program. At its conclusion, statistically and clinically significant changes were noted in weight and BMI measures based on pre-treatment and comparative measurements. The authors of this paper concluded that *this provides supporting evidence on the effectiveness of a multi-modal approach to weight loss implemented in a chiropractic clinic.* The truth, however, is that one could combine any useless intervention with a calorie controlled diet, exercise and one-on-one counselling to create a multi-modal programme for weight loss misleading the public into thinking that the useless intervention was effective.

The claim that 'chiropractic adds years to your life' is yet another popular claim by chiropractors. It seems fair to assume that chiropractors themselves are the best informed about what they perceive as the health benefits of chiropractic care. Chiropractors would therefore be most likely to receive some level of this 'life-prolonging' chiropractic care on a long-term basis. Therefore, chiropractors should, on average, reach an older age than the general population. In 2004, a chiropractor tested this theory and published an interesting paper about it.[17] He used two separate data sources to define the mortality rates of chiropractors. One source used obituary notices from past issues of Dynamic Chiropractic from 1990 to mid-2003. The second source used biographies from 'Who Was Who in Chiropractic—A Necrology' covering a ten-year period from 1969–1979. The two sources yielded a mean age at death for chiropractors of 73.4 and 74.2 years respectively. The mean ages at death of chiropractors thus turned out to be below the national average of 76.9 years; it also is below the average age at death of their medical doctor counterparts which, at the time, was 81.5. So, instead of claiming that 'chiropractic adds years to your life', one might be tempted to conclude that 'chiropractic subtracts years from your life'. Morgan, the author of the paper, concluded *that this paper assumes chiropractors should, more than any other group, be able to demonstrate the health and longevity benefits of chiropractic care. The chiropractic mortality data presented in this study, while limited,*

---

[16] DeMaria A, DeMaria C, Demaria R, Alcantara J. A weight loss program in a chiropractic practice: a retrospective analysis. *Complement Ther Clin Pract.* 2014;20(2):125–129. https://doi.org/10.1016/j.ctcp.2013.11.007.

[17] Morgan L. Does chiropractic 'add years to life'?. *J Can Chiropr Assoc.* 2004;48(3):217–224.

do not support the notion that chiropractic care "Adds Years to Life …", and it fact shows male chiropractors have shorter life spans than their medical doctor counterparts and even the general male population. Further study is recommended to discover what factors might contribute to lowered chiropractic longevity.

Many chiropractors also claim that their spinal manipulation therapy (SMT) enhances athletic performance. But is this claim true? The objective of this study[18] was to systematically review the literature on the effect of SMT on performance-related outcomes in asymptomatic adults. Of 1415 articles screened, 20 studies had low risk of bias, seven were randomized crossover trials, 10 were randomized controlled trials (RCT) and three were RCT pilot trials. The studies showed SMT had no effect on physiological parameters at rest or during exercise. Sport-specific studies showed no effect of SMT except for a small increase in basketball free-throw accuracy. The authors concluded *that the preponderance of evidence suggests that SMT in comparison to sham or other interventions does not enhance performance-based outcomes in asymptomatic adult population.*

Another common claim of chiropractors is that regular chiropractic treatments will improve your quality of life or wellness. A survey showed that 77% of the websites of Australian chiropractors include claims about wellness promotion.[19] Most books on the subject promote it as well; some are even entirely dedicated to the theme.[20] But what is the evidence that chiropractic interventions truly affect quality of life or wellness? Recently, Australian researchers published a review[21] of all 12 investigations of the subject. The results show a high degree of inconsistency. The authors concluded that *it is difficult … to defend any conclusion about the impact of chiropractic intervention on the quality of life …*

It seems almost impossible to find a condition for which some chiropractors do not recommend their services. And sadly, the promotion of false claims is not confined to a small proportion of rogue chiropractors. There is evidence that 95% of chiropractor websites make unsubstantiated claims

---

[18]Corso M, Mior SA, Batley S, et al. The effects of spinal manipulation on performance-related outcomes in healthy asymptomatic adult population: a systematic review of best evidence. *Chiropr Man Therap.* 2019; 27:25. Published 2019 Jun 7. https://doi.org/10.1186/s12998-019-0246-y.

[19]Young KJ. Words matter: the prevalence of chiropractic-specific terminology on Australian chiropractors' websites. *Chiropr Man Therap.* 2020;28(1):18. Published 2020 Apr 7. http://doi.org/10.1186/s12998-020-00306-9.

[20]Feuling TJ. Chiropractic Works. Adjusting to a Higher Quality of Life, 2nd edn, Wellness Solution (2000).

[21]Parkinson L, Sibbritt D, Bolton P, van Rotterdam J, Villadsen I. Well-being outcomes of chiropractic intervention for lower back pain: a systematic review. *Clin Rheumatol.* 2013;32(2):167–180. http://doi.org/10.1007/s10067-012-2116-z.

that have the potential to harm consumers.[22] It is therefore crucial to realise that this plethora of therapeutic claims is not based on sound evidence from clinical research, a theme that will re-emerge in the following chapters.

Box 7.1 Non-spinal conditions which are regularly treated by chiropractors

- Adrenal disorders
- Allergies
- Asthma
- Diabetes
- Ear infections
- Hiatal hernia
- Hypertension
- Immunological dysfunction
- Menopause
- Menstrual disorders
- Nutritional disorders
- Peripheral vascular disease
- Pregnancy related conditions
- Respiratory infections
- Sinus conditions
- Sleep disorders
- Thyroid disorders.

---

[22] Ernst E, Gilbey A. Chiropractic claims in the English-speaking world. *N Z Med J*. 2010;123(1312):36–44. Published 2010 Apr 9.

8

Other Manual Therapies

Manual therapy … is a physical treatment primarily used by physical therapists, physiotherapists to treat musculoskeletal pain and disability; it mostly includes kneading and manipulation of muscles, joint mobilization and joint manipulation. It's also used by occupational therapists, chiropractors, massage therapists, athletic trainers, osteopaths, and physicians. (Wikipedia)

Besides chiropractic, many further manual therapies exist and are used for a wide range of conditions. Many are derived from chiropractic, influenced by chiropractic, or in some ways similar to chiropractic. In this chapter, I will briefly discuss the most important of these treatments.

Bowen Technique

This treatment was developed by the Australian Thomas Ambrose Bowen (1916–1982) who believed that certain movements of the body resulted in particular responses. He thus tried using this observation for treating symptoms. This led to him to develop his own manual therapy and eventually make far-reaching claims for it. Bowen had no previous formal training in any healthcare discipline and, like DD Palmer, stated that his innovation was 'a gift from God'.

The treatment consists of manual mobilisations called 'Bowen moves' over muscles, tendons, nerves and fascia. The therapist performs these rolling movements using the thumbs and fingers applying gentle pressure. A full treatment consists of a standardised sequence of moves, interrupted by regular

© Springer Nature Switzerland AG 2020
E. Ernst, *Chiropractic*,
https://doi.org/10.1007/978-3-030-53118-8_8

pauses to allow time for the body to respond. Patients usually experience this therapy as relaxing and enjoyable.

A 2015 report by the Australian government reviewed the evidence for the Bowen technique and stated that *no clear evidence of effectiveness was found*.[1] There are no rigorous clinical trials testing the effectiveness of the Bowen technique for any disease or symptom. Yet proponents tend to recommend it as a 'cure-all'.

Craniosacral Therapy

Craniosacral therapy (CST sometimes also called 'craniosacral osteopathy') was developed by the US osteopath William Sutherland (1873–1953) and further refined by the US osteopath John Upledger (1932–2012). The treatment consists of gentle touch and palpation of the joints of the skull and sacrum. Practitioners believe that these joints allow enough movement to influence the pulsation of the cerebrospinal fluid which, in turn, improves what they call 'primary respiration'. The notion of 'primary respiration' is based on the following 5 assumptions:

- inherent motility of the central nervous system
- fluctuation of the cerebrospinal fluid
- mobility of the intracranial and intraspinal dural membranes
- mobility of the cranial bones
- involuntary motion of the sacral bones.

Practitioners of CST believe that palpation of the cranium can detect a rhythmic movement of the cranial bones and use gentle pressure to manipulate the cranial bones to achieve a therapeutic result. The degree of mobility and compliance of the cranial bones is, however, minimal and these assumptions lack plausibility.

The therapeutic claims made for CST are not supported by sound evidence. A systematic review of all 6 trials concluded that *the notion that CST is associated with more than non-specific effects is not based on evidence*

---

[1] Document available at www.health.gov.au/internet/main/publishing.nsf/content/0E9129B3574FCA5 3CA257BF0001ACD11/$File/Natural%20Therapies%20Overview%20Report%20Final%20with% 20copyright%2011%20March.pdf.

from rigorous RCTs.[2] Craniosacral therapy is particularly popular for infants. But here too, the evidence fails to show effectiveness. A study concluded that *healthy preterm infants undergoing an intervention with craniosacral therapy showed no significant changes in general movements compared to preterm infants without intervention.*[3]

Dorn Method

The Dorn method is one of the lesser-known manual therapies. Here is some first-hand information about it[4]:

> Developed by Dieter Dorn in the 1970's in the South of Germany … the Method is featured in numerous books and medical expositions, taught to medical students in some universities, covered by most private medical insurances and more and more recognized in general…
>
> As of now only licensed Therapists, Non Medical Practitioners (in Germany called Heilpraktiker (Healing Practitioners with Government recognition) (8), Physical Therapists or Medical Doctors are authorized to practice with government license, but luckily the Dorn Method is mainly a True Self Help Method therefore all other Dorn Method Practitioners can legally help others by sharing it in this way.

> … Every disease, even up to the psychological domain can be treated (positively influenced) unless an illness had already led to irreversible damages at organs. The main areas of application are: Muscle-Skeletal Disorders (incl. Back Pain, Sciatica, Scoliosis, Joint-Pain, Muscular Tensions, Migraines etc.)

There has not been a single clinical trial of the Dorn method. Therefore, it might be wise to avoid this treatment, particularly as it is one of the more expensive manual therapies on offer.

---

[2]Ernst, E. (2012), Review. Focus on Alternative and Complementary Therapies, 17: 197–201. https://doi.org/10.1111/j.2042-7166.2012.01174.x.

[3]Raith W, Marschik PB, Sommer C, et al. General Movements in preterm infants undergoing craniosacral therapy: a randomised controlled pilot-trial. *BMC Complement Altern Med*. 2016;16:12. Published 2016 Jan 13. https://doi.org/10.1186/s12906-016-0984-5.

[4]http://www.dorn-method.com/dornmethod_introduction.html.

Khalifa Therapy

Khalifa therapy is named after the Australian practitioner, Mohamed Khalifa, who invented this treatment. It consists of the rhythmic application of manual pressure on parts of the patient's body. Which allegedly stimulates the self-healing processes of the human body.

Khalifa therapy has been tested in one small clinical trial. It concluded that this treatment can be helpful in the repair of anterior cruciate ligament ruptures.[5] However, without an independent replication, these findings cannot be accepted. The treatment seems not to be associated with significant direct risks.

Naprapathy

Oakley Smith was a former Iowa medical student who also investigated Andrew Still's osteopathy before becoming a follower of DD Palmer in Davenport (Chap. 2). Smith later came to reject Palmer's concept of vertebral subluxation (Chap. 4) and developed his own "connective tissue doctrine" which later became known as naprapathy. Today, it is a popular form of manual therapy, particularly in Scandinavia and the US, which is being promoted for a very wide range of conditions.

The nature of Smith's approach has been explained as follows[6]:

Naprapathy is defined as a system of specific examination, diagnostics, manual treatment and rehabilitation of pain and dysfunction in the neuromusculoskeletal system. The therapy is aimed at restoring function through treatment of the connective tissue, muscle- and neural tissues within or surrounding the spine and other joints. Naprapathic treatment consists of combinations of manual techniques for instance spinal manipulation and mobilization, neural mobilization and Naprapathic soft tissue techniques, in additional to the manual techniques Naprapaths uses different types of electrotherapy, such as ultrasound, radial shockwave therapy and TENS. The manual techniques are often combined with advice regarding physical activity and ergonomics as well as medical rehabilitation training in order to decrease pain and disability and increase work ability and quality of life. A Dr. of Naprapathy is specialized

---

[5]Ofner M, Kastner A, Wallenboeck E, et al. Manual khalifa therapy improves functional and morphological outcome of patients with anterior cruciate ligament rupture in the knee: a randomized controlled trial. *Evid Based Complement Alternat Med.* 2014;2014:462840. https://doi.org/10.1155/2014/462840.

[6]https://edzardernst.com/2018/02/naprapathy-a-lot-of-it-looks-just-like-quackery-to-me/.

in the diagnosis of structural and functional neuromusculoskeletal disorders, treatment and rehabilitation of patients with problems of such origin as well as to differentiate pain of other origin.

The US 'National College of Naprapathic Medicine' grants the title 'Doctor of Naprapathy' (D.N.). There are three clinical trials of naprapathy listed on Medline. They do not provide good evidence that naprapathy is effective or superior to other therapeutic options.

Osteopathy

Osteopathy is a form of manual therapy invented by the American Andrew Taylor Still (1828–1917) several years before Palmer developed chiropractic. There is evidence that Palmer 'borrowed' many of his ideas from Still (Chap. 2). Yet, Palmer was repeatedly scathing about osteopathy. Today, there are two very different types of osteopaths (Fig. 8.1):

- US osteopaths (doctors of osteopathy or DOs) have more or less stopped practising manual therapy; they are fully recognised as medical doctors and can specialise in any medical field after their training which is almost identical to that of MDs.
- Outside the US, osteopaths practice almost exclusively manual treatments and are considered alternative practitioners. This discussion deals with the latter category of osteopaths.

Andrew Still defined his original osteopathy as a *science which consists of such exact, exhaustive, and verifiable knowledge of the structure and function of the human mechanism, anatomical, physiological and psychological, including the chemistry and physics of its known elements, as has made discoverable certain organic laws and remedial resources, within the body itself, by which nature under the scientific treatment peculiar to osteopathic practice, apart from all ordinary methods of extraneous, artificial, or medicinal stimulation, and in harmonious accord with its own mechanical principles, molecular activities, and metabolic processes, may recover from displacements, disorganizations, derangements, and consequent disease, and regained its normal equilibrium of form and function in health and strength.*[7]

---

[7] Still AT. Philosophy of Osteopathy. CreateSpace Independent Publishing (2016).

Fig. 1 Andrew Taylor Still; *Source* US National Library of Medicine

A more recent definition was published on the website of the London-based 'University College of Osteopathy' (UCO)[8]:

> Osteopathy is a person-centred manual therapy that aims to enable patients to respond and adapt to changing circumstances and to live well... osteopathy has the potential to help people change their lives – not only by searching for ways to manage disease, but also by helping patients to discover ways to enhance and maintain their own health and wellbeing.
>
> A core principle of osteopathy is that wellbeing is dependent on how each person is able to function and adapt to changes in physical capability and their environment. Osteopaths are often described as treating the individual rather than the condition: when treating a patient they consider the symptom or injury alongside other biological, physiological and social factors which may be contributing to it.
>
> Osteopaths work to ensure the best possible care for their patients, aiding their recovery and supporting them to help manage their conditions through a range of approaches, including physical manipulation of the musculoskeletal system and education and advice on exercise, diet and lifestyle.

One important difference between osteopathy and chiropractic is that osteopaths tend to use less HVLA thrusts which have been associated with serious adverse effects (Chap. 14). This means that osteopathy has less potential to cause serious harm than chiropractic.

An overview of 100 randomly selected websites of osteopaths revealed that 93% checked at least one of the criteria for pseudo-scientific claims. The author concluded that *quackery is rife in osteopathic practice.*[9] Some osteopaths consider themselves as back pain specialists, while others claim to effectively treat a much wider range of conditions. The evidence for osteopathy is disappointing.

- For back pain, the evidence is encouraging but not conclusively positive. One review (by osteopaths) concluded that osteopathic treatment *significantly reduces low back pain. The level of pain reduction is greater than expected from placebo effects alone and persists for at least three months. Additional research is warranted to elucidate mechanistically how osteopathic manipulative exerts its effects, to determine if OMT benefits are long lasting,*

---

[8]https://www.uco.ac.uk/about-osteopathy/what-osteopathy.

[9]https://appletzara.wordpress.com/2016/04/24/osteopathy-part-2-a-review-of-100-osteopathy-web sites/.

and to assess the cost-effectiveness of osteopathic manipulative treatment as a complementary treatment for low back pain.[10]

- An independent review, however, found that the *data fail to produce compelling evidence for the effectiveness of osteopathy as a treatment of musculoskeletal pain.*[11]
- For non-spinal conditions, the evidence is even less convincing. One review concluded, for instance, that *the evidence of the effectiveness of osteopathic manipulative therapy for pediatric conditions remains unproven due to the paucity and low methodological quality of the primary studies.*[12]

Even though adverse effects after osteopathy are less frequent than with chiropractic treatments, severe complications have been noted and *include cauda equina syndrome, lumbar disk herniation, fracture, and hematoma or haemorrhagic cyst. Contraindications … primarily involve conditions that increase bleeding risk or compromise bone, tendon, ligament, or joint integrity.*[13]

Polarity Therapy

Polarity therapy is an approach that combines bodywork, diet, exercise and lifestyle counselling. It was invented by the Austrian-American chiropractor, osteopath, and naturopath Randolph Stone (1888–1981) who aimed at an integration of Eastern and Western philosophies, principles and techniques of healing. Similar to chiropractic, polarity therapy is based on the assumption that our health depends on the energy flow within our body and that polarity therapy can re-adjust this flow when necessary.

Polarity therapy is claimed to work for a wide range of conditions, including intoxications, HIV, stress, back pain, and stomach cramps. After determining the assumed source of a patient's energy imbalance, the therapist begins the first of a series of bodywork sessions designed to re-channel and release the patient's misdirected energy. This may be followed by a series of

---

[10]Licciardone JC, Brimhall AK, King LN. Osteopathic manipulative treatment for low back pain: a systematic review and meta-analysis of randomized controlled trials. *BMC Musculoskelet Disord.* 2005;6:43. Published 2005 Aug 4. https://doi.org/10.1186/1471-2474-6-43.

[11]Posadzki P, Ernst E. Osteopathy for musculoskeletal pain patients: a systematic review of randomized controlled trials. *Clin Rheumatol.* 2011;30(2):285–291. https://doi.org/10.1007/s10067-010-1600-6.

[12]Posadzki P, Lee MS, Ernst E. Osteopathic manipulative treatment for pediatric conditions: a systematic review. *Pediatrics.* 2013;132(1):140–152. https://doi.org/10.1542/peds.2012-3959.

[13]Jonas C. Musculoskeletal Therapies: Osteopathic Manipulative Treatment. *FP Essent.* 2018;470:11-15.

exercises called polarity yoga which include squats, stretches, rhythmic movements, deep breathing, and expression of sounds. They are assumed to be energizing and relaxing. Counselling may be included whenever appropriate as a part of a patient's highly individualised therapy regimen to promote balance.

According to its proponents, polarity therapy *addresses many different levels: subtle energy; nervous, musculo-skeletal, cardiovascular, myofascial, respiratory and digestive systems; as well as the emotional and mental levels. It tackles many varied and different expressions of disease by unlocking the holding patterns that create the symptoms. Dr. Stone called it Polarity because it embodies the negative and positive poles and the neutrality between them. This Modality is a healing art, an anchor for a balanced lifestyle, an evolving way of being, an adjunct to other modalities.*[14]

There have been only very few clinical trials testing the effectiveness of polarity therapy, and those that have emerged are less than rigorous. Therefore, the effectiveness of polarity therapy is unproven.

Rolfing

Rolfing is a manual therapy invented by Ida Pauline Rolf (1896–1979); it employs deep manipulation of the body's soft tissue allegedly to realign and balance the body's myofascial structures. The 'Guild for Structural Integration' describe Rolfing as *a method and a philosophy of personal growth and integrity…. The vertical line is our fundamental concept. The physical and psychological embodiment of the vertical line is a way of Being in the physical world [that] forms a basis for personal growth and integrity.*[15]

Rolfing is being promoted as a system that reshapes the body's myofascial structure by applying pressure and energy. It is said to free the body from the effects of physical and emotional traumas. A key assumption is that fascia tightens from chronic dysfunctional movement and imbalanced muscular tension, resulting in contracture. Over time, the tightened fascia becomes shortened with disorganized collagen, constricting muscle contraction and relaxation, and consequently constricting the movements of our joints. Rolf stated that *Rolfers make a life study of relating bodies and their fields to the earth and its gravity field, and we so organize the body that the gravity field can reinforce the body's energy field. This is our primary concept.*[16] Proponents

---

[14]http://www.polaritytherapy.org.uk/.

[15]https://www.rolfguild.org/mission.

[16]Rolf IP. Rolfing and Physical Reality. Inner Traditions/Bear (1990).

claim that Rolfing can bring relief from chronic back, neck, shoulder and joint pain, improve breathing, increase energy, improve self-confidence, and relieve physical and mental stress.

Only very few trials of Rolfing have emerged, and none are of good quality.[17] Therefore, the therapeutic claims made for Rolfing are not evidence-based. Rolfing involves vigorous deep, somewhat forceful tissue manipulation and is often experienced as uncomfortable or painful.

Tragerwork

The US doctor Milton Trager developed 'Tragerwork' in the 1970s. Milton Trager had been a boxer who became involved in bodywork therapy when he realised that he was able to feel people's tensions in their body and relieve them with gentle movements. He then trained as a physical therapist and later also qualified as a doctor.

Tragerwork is claimed to release deleterious, so-called "holding patterns" allegedly found in muscles through gentle, rocking massage and to enhance the interaction between the body and the mind. Tragerwork therapists use their hands and minds to communicate feelings of lightness and freedom to their patients. Aided movements are followed by lessons in 'mentastics' (mental gymnastics) involving dance-like movements which are said to enhance a sensation of lightness.

The therapeutic claims for Tragerwork are often far-reaching[18] and include the assumptions of effectively treating asthma, autism, depression, emphysema, hypertension, low back pain, migraines, multiple sclerosis, muscular dystrophy, polio, neuromuscular diseases, pain, poor posture, sciatica, sports injuries, improvement of athletic performance, mental control, responsiveness and conservation of energy in movement.

There have been only very few clinical trials testing the effectiveness of Tragerwork. Those that have emerged are methodologically too weak to allow meaningful conclusions.[19] Thus, none of the many therapeutic claims made by proponents of this therapy are supported by sound evidence.

---

[17] Jones TA. Rolfing. Phys Med Rehabil Clin N Am 15 (2004) 799–809.

[18] Russell JK. Bodywork--the art of touch. *Nurse Pract Forum*. 1994;5(2):85–90.

[19] Mehling WE, DiBlasi Z, Hecht F. Bias control in trials of bodywork: a review of methodological issues. *J Altern Complement Med*. 2005;11(2):333–342. https://doi.org/10.1089/acm.2005.11.333.

Visceral Osteopathy

Visceral osteopathy (or visceral manipulation) is an expansion of the principles of osteopathy and involves the manual manipulation of internal organs, blood vessels and nerves (the viscera) from outside the body. Visceral osteopathy was developed by the osteopath Jean-Pierre Barral. He stated that *through his clinical work with thousands of patients, he created this modality based on organ-specific fascial mobilization. And through work in a dissection lab, he was able to experiment with VM techniques and see the internal effects of the manipulations.*[20]

According to its proponents, visceral manipulation *is based on the specific placement of soft manual forces looking to encourage the normal mobility, tone and motion of the viscera and their connective tissues. These gentle manipulations may potentially improve the functioning of individual organs, the systems the organs function within, and the structural integrity of the entire body.*[21]

Visceral osteopathy is used by osteopaths, chiropractors and physiotherapists. It comprises a range of different manual techniques firstly for diagnosing a health problem and secondly for treating it. Several studies have assessed the diagnostic reliability of the techniques involved. The totality of this evidence fails to show that they are valid.[22] Other studies have tested whether the therapeutic techniques used in visceral osteopathy are effective in curing disease or alleviating symptoms. The totality of this evidence fails to show that visceral osteopathy works for any condition.

The treatment itself is usually safe, yet the risks of visceral osteopathy are nevertheless considerable: if a patient suffers from symptoms related to her inner organs, a visceral osteopath is likely to misdiagnose them and subsequently mistreat them. If the symptoms are due to a serious disease, this would amount to medical neglect and could, in extreme cases, cost the patient's life.

---

[20]https://www.barralinstitute.com/about/jean-pierre-barral.php.

[21]http://www.barralinstitute.co.uk/.

[22]Guillaud A, Darbois N, Monvoisin R, et al. Reliability of diagnosis and clinical efficacy of visceral osteopathy: a systematic review. *BMC Complement Altern Med* **18,** 65 (2018). https://doi.org/10.1186/s12906-018-2098-8.

9

What Is Evidence?

We promote evidence-based practice[1] (World Federation of Chiropractic)

Evidence is clearly not in line with most of the therapeutic claims made by chiropractors (Chaps. 10–12). Yet, few chiropractors seem worried about this fact; they tend to believe that their daily experience (and that of thousands of their colleagues as well as millions of patients) holds information about the effectiveness of their spinal manipulations. What is more, they feel that their experience is more reliable than any scientific investigation. Whenever their patients get better, they assume this to be the result of their spinal adjustments. But this conclusion is not necessarily warranted; the reason is simple: two events—the treatment by the chiropractor and the improvement of the patient—that follow each other in time are not necessarily causally related. In fact, there are several other factors that determine the outcome, for instance:

- the natural history of the condition,
- regression towards the mean,
- the placebo effect.

Consequently, ineffective or even mildly harmful treatments can appear to be effective. If we want to know whether a treatment really works, we need to control for all the factors that might confound the outcome. Ideally, we need an experiment where two groups of patients are exposed to the full range of

---

[1] https://www.wfc.org/website/index.php?option=com_content&view=article&id=533:wfc-releases-new-guiding-principles-document&catid=56:news--publications&Itemid=27&lang=en&fbclid=IwAR26GqFjENyJEfGD9_Pseoo-9jcnX8UlrFKXiyFuIOEUWPbqThO5IL_yvew.

© Springer Nature Switzerland AG 2020
E. Ernst, *Chiropractic*,
https://doi.org/10.1007/978-3-030-53118-8_9

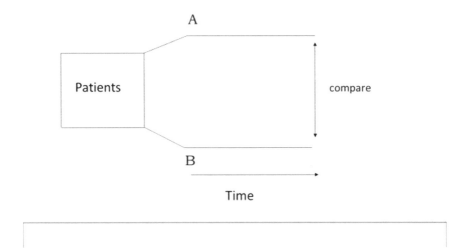

Fig. 9.1 Schematic explanation of the principle of a controlled clinical trial

confounding factors, and the only difference between them is that one group does receive the treatment, while the other group does not.

This is precisely the concept of a controlled clinical trial. This methodology accounts for all the factors which otherwise might cloud our judgement. In a typical controlled clinical trial of chiropractic manipulation, researchers divide a group of patients into one subgroup (A) who receive the treatment and a second subgroup (B), called the control group. The control group receives a different treatment, for instance, a sham manipulation or a conventional therapy (the exact choice depends on the precise research question). The two treatments are then administered for the proscribed length of time. At the end of this period, the results of the two groups are compared. If the chiropractic group demonstrates superior outcomes to the control group, the former is deemed to be effective (Fig. 9.1).

Here is an interesting example of one of the few clinical trials comparing spinal manipulation with sham manipulation, i.e. placebo controls[2]:

The purpose of this study was to compare the effectiveness of chiropractic spinal manipulative therapy (cSMT) to a sham intervention on pain (Visual

---

[2]Dougherty PE, Karuza J, Dunn AS, Savino D, Katz P. Spinal Manipulative Therapy for Chronic Lower Back Pain in Older Veterans: A Prospective, Randomized, Placebo-Controlled Trial. *Geriatr Orthop Surg Rehabil*. 2014;5(4):154-164. http://doi.org/10.1177/2151458514544956.

Analogue Scale, SF-36 pain subscale), disability (Oswestry Disability Index), and physical function (SF-36 subscale, Timed Up and Go) by performing a randomized placebo-controlled trial at 2 Veteran Affairs Clinics. Older veterans ($\geq$ 65 years of age) who were naive to chiropractic were recruited. A total of 136 who suffered from chronic low back pain (LBP) were included in the study – with 69 being randomly assigned to cSMT and 67 to the sham intervention. Patients were treated twice per week for 4 weeks. The outcomes were assessed at baseline, 5, and 12 weeks post baseline. Both groups demonstrated significant decrease in pain and disability at 5 and 12 weeks. At 12 weeks, there was no significant difference in pain and a statistically significant decline in disability scores in the cSMT group when compared to the control group. There were no significant differences in adverse events between the groups. The authors concluded that *cSMT did not result in greater improvement in pain when compared to our sham intervention; however, cSMT did demonstrate a slightly greater improvement in disability at 12 weeks. The fact that patients in both groups showed improvements suggests the presence of a nonspecific therapeutic effect.*

Many different variations of the controlled trial exist, and the exact design can be adapted to the requirements of the specific research question at hand. The over-riding aim, however, is always the same: the investigators want to make sure that they can reliably determine whether the treatment was the cause of the observed outcome.

Causality is, of course, the key in all of this; and it is the main difference between clinical experience and scientific evidence. The outcomes chiropractors observe and their patients note can have a myriad of causes. By contrast, what scientists register in a well-designed trial is caused by the treatment. The latter is evidence, while the former is not.

Sadly, clinical trials are rarely perfect; they can have numerous limitations, and even those funded by the most prestigious sources can be flawed. In 2011, we published an evaluation of all the randomised clinical trials (RCTs) of chiropractic funded by the National Centre for Complementary and Alternative Medicine (NCCAM) of the US National Institutes of Health.[3] Ten RCTs were included, mostly related to chiropractic SMT for musculoskeletal problems. Their quality was frequently questionable. Several RCTs failed to report adverse effects (which means they were violating ethical standards) and the majority was not described in sufficient detail to allow replication. We concluded that *the criticism repeatedly aimed at NCCAM seems justified, as far*

---

[3] Ernst E, Posadzki P. An independent review of NCCAM-funded studies of chiropractic. *Clin Rheumatol*. 2011;30(5):593–600. http://doi.org/10.1007/s10067-010-1663-4.

as their RCTs of chiropractic is concerned. It seems questionable whether such research is worthwhile.

But, despite all their potential shortcomings, well-conducted clinical trials are far superior to any other currently known method for determining the effectiveness of medical interventions. And, because clinical trials can occasionally be unreliable, we should avoid relying on the findings of one single study. Independent replications are usually required to be sure. But often the findings of such replications do not confirm the results of the previous study.

Whenever we are faced with conflicting results, it is tempting to cherry-pick those studies which confirm our prior belief—tempting but wrong. But sadly, this wrong approach is often taken in the realm of chiropractic. An example is this book entitled 'Practice Analysis of Chiropractic'. Its chapter on 'Chiropractic Research'[4] summarises the evidence for chiropractic, citing surveys, observational studies, clinical trials and reviews—but only those that arrive at conclusions favoured by the 'National Board of Chiropractic Examiners' (NBCE). This means that negative findings are carefully eliminated. What is particularly impressive is the fact that the NBCE implies that chiropractic SMT is helpful for a wide range of non-spinal paediatric conditions. Why does the NBCE perpetuate the myth? The best answer I can find is that does it to support chiropractors' income.

In order to arrive at the most reliable conclusion, we need to consider the totality of the reliable evidence. This goal is best achieved by conducting a systematic review (in this book, I therefore rely on this type of evidence whenever possible).

In a systematic review, we assess the quality and quantity of the available evidence, synthesise the findings, and arrive at an overall verdict. Technically speaking, this process minimises both selection and random biases. Systematic reviews (and meta-analyses, i.e. systematic reviews that pool the data of individual studies and calculate a new quantitative result) therefore constitute the best available evidence for or against the effectiveness of any treatment, including of course chiropractic.

But even with systematic reviews, it is advisable to be critical. Today, there are dozens of reviews of chiropractic, and not all are reliable. We have assessed the possible sources of bias in such publications and found that their quality was hugely variable. Those of poor quality tended to be authored by chiropractors and reported positive findings, while the rigorous reviews failed to do so. We concluded that *the outcomes of reviews of this subject are strongly influenced by both scientific rigour and profession of authors. The effectiveness of*

---

[4]Document available at https://chiro.org/LINKS/FULL/Practice_Analysis_of_Chiropractic_2015/Chapter_2_Review_of_Research_Literature.pdf.

spinal manipulation for back pain is less certain than many reviews suggest; most high quality reviews reach negative conclusions.[5] The technically best and most independent systematic reviews are usually those published by the Cochrane Collaboration (Box 9.1).

Only with reliable evidence can we tell with any degree of certainty that it was the chiropractic treatment—and not any of the other factors mentioned above—that caused the clinical outcome we observe in a group of patients. Only if we have such evidence can we be sure about cause and effect. And only then can we make sure that patients receive the best possible treatments currently available.

Box 9.1 Conclusions of four recent Cochrane reviews of spinal manipulation

- **Neck pain**: Although support can be found for use of thoracic manipulation versus control for neck pain, function and QoL, results for cervical manipulation and mobilisation versus control are few and diverse. Publication bias cannot be ruled out. Research designed to protect against various biases is needed. Findings suggest that manipulation and mobilisation present similar results for every outcome at immediate/short/intermediate-term follow-up. Multiple cervical manipulation sessions may provide better pain relief and functional improvement than certain medications at immediate/intermediate/long-term follow-up. Since the risk of rare but serious adverse events for manipulation exists, further high-quality research focusing on mobilisation and comparing mobilisation or manipulation versus other treatment options is needed to guide clinicians in their optimal treatment choices.
- **Acute low back pain**: SMT is no more effective in participants with acute low-back pain than inert interventions, sham SMT, or when added to another intervention. SMT also appears to be no better than other recommended therapies. Our evaluation is limited by the small number of studies per comparison, outcome, and time interval. Therefore, future research is likely to have an important impact on these estimates. The decision to refer patients for SMT should be based upon costs, preferences of the patients and providers, and relative safety of SMT compared to other treatment options. Future RCTs should examine specific subgroups and include an economic evaluation.
- **Chronic low back pain**: High quality evidence suggests that there is no clinically relevant difference between SMT and other interventions for reducing pain and improving function in patients with chronic low-back pain. Determining cost-effectiveness of care has high priority. Further research is likely to have an important impact on our confidence in the estimate of effect in relation to inert interventions and sham SMT, and data related to recovery.

---

[5]Canter PH, Ernst E. Sources of bias in reviews of spinal manipulation for back pain. *Wien Klin Wochenschr.* 2005;117(9-10):333–341. http://doi.org/10.1007/s00508-005-0355-6.

- **Infant colic:** The studies included in this meta-analysis were generally small and methodologically prone to bias, which makes it impossible to arrive at a definitive conclusion about the effectiveness of manipulative therapies for infantile colic. The majority of the included trials appeared to indicate that the parents of infants receiving manipulative therapies reported fewer hours crying per day than parents whose infants did not, based on contemporaneous crying diaries, and this difference was statistically significant. The trials also indicate that a greater proportion of those parents reported improvements that were clinically significant. However, most studies had a high risk of performance bias due to the fact that the assessors (parents) were not blind to who had received the intervention. When combining only those trials with a low risk of such performance bias, the results did not reach statistical significance. Further research is required where those assessing the treatment outcomes do not know whether or not the infant has received a manipulative therapy. There are inadequate data to reach any definitive conclusions about the safety of these interventions.

10

Effectiveness of Spinal Manipulation for Spinal Problems

Chiropractors specialise in assessing, diagnosing and managing conditions of the spine.
—British Chiropractic Association

Almost all of us will have low back pain at some point in our lives. The problem can affect anyone at any age, and it seems to be increasing: disability due to back pain has risen by more than 50% since 1990.[1] Treatment of back pain varies widely (Box 10.1); the options offered range from bed rest to surgery, from yoga to the use of painkillers, from massage to chiropractic.

Spinal manipulation is the hallmark therapy of chiropractors (Chap. 5), and back pain is the condition chiropractors treat more often than any other (Chap. 7). This clearly begs the question, is chiropractic spinal manipulation therapy (SMT) effective for back pain? Unsurprisingly, chiropractors try to make us believe that it is. However, as we shall discuss in this chapter, the evidence is far from convincing.

---

[1] https://www.thelancet.com/series/low-back-pain.

© Springer Nature Switzerland AG 2020
E. Ernst, *Chiropractic*,
https://doi.org/10.1007/978-3-030-53118-8_10

Treatment of Acute Low Back Pain

Numerous clinical trials have tested the effectiveness of SMT for back pain. Their findings are hugely contradictory. Thus, chiropractors are able to cherry-pick those results that suit them, while critics might select those studies that came out negative. Neither approach would be fair or constructive. The only reliable way forward in such a situation is to assess the totality of the robust evidence (Chap. 9); in other words, we need to look at high quality systematic reviews published by independent researchers.

The aim of this systematic review was to evaluate all studies of the effectiveness of SMT for acute low back pain.[2] Of 26 eligible randomised clinical trials (RCTs) identified, 15 RCTs (with a total of 1711 patients) provided moderate-quality evidence that SMT reduces back pain. Twelve RCTs produced moderate-quality evidence that SMT improves function. Minor transient adverse events such as increased pain, muscle stiffness, and headache were reported 50–67% of the time in large case series of patients treated with SMT. The authors concluded that *among patients with acute low back pain, spinal manipulative therapy was associated with modest improvements in pain and function at up to 6 weeks, with transient minor musculoskeletal harms. However, heterogeneity in study results was large.*

This paper was celebrated by chiropractors as a proof that their SMT works. However, we should view the findings in perspective: the effect size was small and would easily be achievable by other, less harmful and cheaper therapies. One critic went further[3]:

> … the review itself does seem positive at first glance: the benefits of SMT are disingenuously summarized as "statistically significant" in the abstract, with no mention of clinical significance. So the abstract sounds like good news to anyone but the most wary readers, while deep in the main text the same results are eventually conceded to be "clinically modest." … SMT is "just" disappointingly mediocre on average, but might have more potent benefits in a minority of cases (that no one seems to be able to reliably identify). Far from being good news, this review continues a strong trend of damning SMT with faint praise, and also adds evidence of backfiring to mix. Although fortunately "no RCT reported any serious adverse event," it seems that minor harms were legion: "increased pain, muscle stiffness, and headache were reported 50% to

---

[2]Paige NM, Miake-Lye IM, Booth MS, et al. Association of Spinal Manipulative Therapy With Clinical Benefit and Harm for Acute Low Back Pain: Systematic Review and Meta-analysis. *JAMA.* 2017;317(14):1451–1460. https://doi.org/10.1001/jama.2017.3086.

[3]https://www.painscience.com/biblio/spinal-manipulation-effects-on-acute-back-pain-range-from-from-negative-to-minor.html.

67% of the time in large case series of patients treated with SMT." That's a lot of undesirable outcomes. So the average patient has a roughly fifty-fifty chance of up to roughly maybe a 20% improvement... or feeling worse to some unknown degree! That does not sound like a good deal to me. It certainly doesn't sound like good medicine.

To this, I would add that none of the studies summarised in the review were controlled for placebo effects. It is therefore possible—I would even say likely—that a sizable part of the observed benefit is not due to SMT per se but to a placebo response.

Treatment of Chronic Low Back Pain

A Cochrane review (Cochrane reviews are considered to be amongst the best) assessed the benefits and harms of SMT for the treatment of chronic low back pain.[4] The authors included all RCTs examining the effect of spinal manipulation or mobilisation in adults with chronic low back pain with or without referred pain. Studies that exclusively examined sciatica were excluded. The effect of SMT was compared with recommended therapies, non-recommended therapies, sham (placebo) SMT, and SMT as an adjuvant therapy. Forty-seven RCTs including a total of 9211 participants were identified (Fig. 10.1). Most trials compared SMT with recommended therapies. High velocity manipulations were used in 18 RCTs, low velocity manipulations in 12 studies and a combination of the two in 20 trials. Moderate quality evidence suggested that SMT has similar effects to other recommended therapies on short term pain function. High quality evidence suggested that, compared with non-recommended therapies, SMT resulted in small, not clinically better effects for short term pain relief and small to moderate clinically better improvement in function. In general, these results were similar for the intermediate and long term outcomes as were the effects of SMT as an adjuvant therapy. Low quality evidence suggested that SMT does not result in a statistically better effect than sham SMT at one month. Additionally, very low quality evidence suggested that SMT does not result in a statistically better effect than sham SMT at six and 12 months. Low quality evidence suggested that SMT results in a moderate to strong statistically significant and clinically better effect than sham SMT at one month.

[4]Rubinstein SM, de Zoete A, van Middelkoop M, Assendelft WJJ, de Boer MR, van Tulder MW, et al. Benefits and harms of spinal manipulative therapy for the treatment of chronic low back pain: systematic review and meta-analysis of randomised controlled trials *BMJ* 2019; 364:l689.

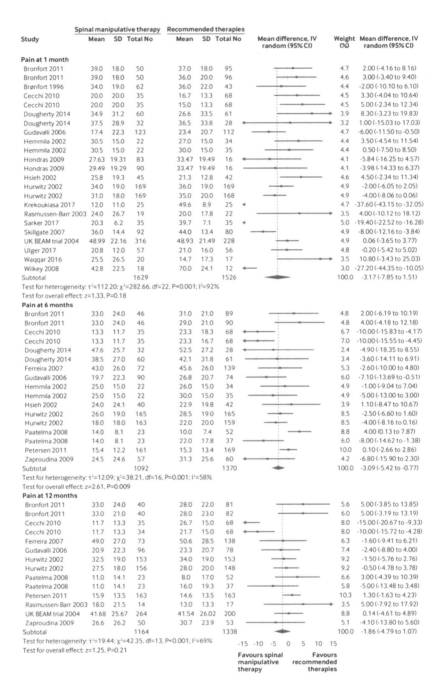

Fig. 10.1 Forest plot summarising all the studies visually. Mean difference in reduction of pain at 1, 3, 6, and 12 months (0–100; 0 = no pain, 100 maximum pain) for spinal manipulative therapy (SMT) versus recommended therapies in reviews of the effects of SMT for chronic low back pain

Additionally, very low quality evidence suggested that SMT does not result in a statistically significant better effect than sham SMT at six and 12 months.

The authors concluded that *SMT produces similar effects to recommended therapies for chronic low back pain, whereas SMT seems to be better than non-recommended interventions for improvement in function in the short term. Clinicians should inform their patients of the potential risks of adverse events associated with SMT.*

Chiropractors tend to believe that this review vindicates their treatments as being both effective and safe. However, this interpretation is clearly over-optimistic:

- SMT is as good as other recommended treatments for back problems. As no good treatment for chronic back pain has yet been identified, this really means that SMT is as bad as other recommended therapies.
- If we have a handful of equally unsatisfactory treatments, it stands to reason that we use criteria other than benefit to identify the one that is best suited—criteria like safety and cost, for instance. If we do that, it becomes clear that SMT is not the treatment of choice (Chap. 14).
- Less than half the RCTs mentioned adverse effects. This means that these studies violated ethical research standards. It is therefore debatable whether we can trust such deeply flawed trials.
- Only 10% of the included RCTs attempted to blind patients to the assigned intervention by providing a sham treatment. Therefore, the observed outcomes are likely to be largely due to the undoubtedly large placebo effects of SMT.

After carefully considering this review, the best conclusion we can draw from it is this: SMT as a treatment for chronic back pain is not supported by convincing evidence.

A series of articles published 2018 in the Lancet confirms this view.[5] In the table below, I have listed the non-pharmacological, non-operative treatments evaluated in the series together with the authors' verdicts regarding their effectiveness for both acute and persistent low back pain (LBP). The Lancet focussed on the effectiveness of several therapeutic options. But the value of a treatment is not only determined by its effectiveness. Crucial further factors are a therapy's cost and its risks, the latter of which also determines the most

---

[5] https://www.thelancet.com/series/low-back-pain.

important criterion: the risk/benefit balance. In my version of the table, I have therefore added these three factors for non-pharmacological and non-surgical options:

	Effectiveness acute LBP	Effectiveness persistent LBP	Risks	Costs	Risk/Benefit Balance
Advice to stay active	+, routine	+, routine	None	Low	Positive
Education	+, routine	+, routine	None	Low	Positive
Superficial heat	±	le	Very minor	Low to medium	Positive (aLBP)
Exercise	Limited	± , routine	Very minor	Low	Positive (pLBP)
CBT	Limited	± , routine	None	Low to medium	Positive (pLBP)
Spinal manipulation	±	±	vfbmae sae	High	Negative
Massage	±	±	Very minor	High	Positive
Acupuncture	±	±	sae	High	Questionable
Yoga	le	±	Minor	Medium	Questionable
Mindfulness	le	±	Minor	Medium	Questionable
Rehab	le	±	Minor	Medium to high	Questionable

Routine = consider for routine use
± = second line or adjunctive treatment
le = insufficient evidence
Limited = limited use in selected patients
vfbmae = very frequent, minor adverse effects
sae = serious adverse effects, including deaths, are on record
aLBP = acute low back pain
pLBP = persistent low back pain
CBT = cognitive behavioural therapy

The Table shows that an optimal treatment for LBP has not yet been identified. In this situation, we should opt for a treatment (amongst similarly effective/ineffective therapies) that is at least safe, cheap and readily available. This treatment is not SMT but it consists of education, staying active and using therapeutic exercise.

A most comprehensive review by the US 'Agency for Healthcare Research and Quality' arrived at similarly negative conclusions: *There was no difference between spinal manipulation versus sham manipulation, usual care, an attention control, or a placebo intervention in short-term.*[6] The authors also found that the strength of the evidence was low.

Another meta-analysis compared the effectiveness of 4 different approaches to treating acute and chronic LBP more directly:

- SMT,
- medical management,
- physical therapies,
- exercise.[7]

The results of this comparison suggest that no treatment is better than sham or placebo. The authors concluded: *Meta-analyses can extract comparative effectiveness information from existing literature. The relatively small portion of outcomes attributable to treatment explains why past research results fail to converge on stable estimates. The probability of treatment superiority between treatment arms was equivalent to that expected by random selection. Treatments serve to motivate, reassure, and calibrate patient expectations—features that might reduce medicalization and augment self-care. Exercise with authoritative support is an effective strategy for acute and chronic low back pain.* This confirms that none of the treatments for low back pain are convincingly effective. We should therefore opt for the therapy with the least risks and costs. The treatment of choice is therefore exercise.

But what type of exercise? The authors of a high-quality review concluded that, *unless supplementary high-quality studies provide different evidence, walking, which is easy to perform and highly accessible, can be recommended in the management of chronic LBP to reduce pain and disability.*[8] A systematic review (authored by a chiropractor and published in a chiropractic journal) seemed to agree with this view when concluding that *"there is no conclusive evidence that clearly favours spinal manipulation or exercise as more effective in*

---

[6]Skelly AC, Chou R, Dettori JR, et al. Noninvasive Nonpharmacological Treatment for Chronic Pain: A Systematic Review Update [Internet]. Rockville (MD): Agency for Healthcare Research and Quality (US); 2020 Apr. (Comparative Effectiveness Review, No. 227.) Evidence Summary. Available from: https://www.ncbi.nlm.nih.gov/books/NBK556236/.

[7]Menke JM. Do manual therapies help low back pain? A comparative effectiveness meta-analysis. *Spine (Phila Pa 1976).* 2014;39(7):E463–E472. https://doi.org/10.1097/BRS.0000000000000230.

[8]Sitthipornvorakul, E. The effects of walking intervention in patients with chronic low back pain: A meta-analysis of randomized controlled trials. Musculoskeletal Science & Practice, Vol 34, 38–46 (2018).

treatment of chronic low back pain."[9] And the new guideline by NICE recommends various forms of exercise as the first step in managing low back pain[10]: *Massage and manipulation … should only be used alongside exercise; there is not enough evidence to show they are of benefit when used alone. Moreover, patients should be encouraged to continue with normal activities as far as possible.*

Treatment of Neck Pain

Neck pain is the second most common condition treated by chiropractors. Here too, the evidence is highly contradictory, and therefore it is mandatory to not rely on cherry-picked trials but on the totality of the robust evidence.

My own systematic review of 2003 found no good evidence to suggest that chiropractic SMT is an effective treatment for neck pain. Here is its abstract[11]:

Chiropractic spinal manipulation (SMT) is often used as a treatment for neck pain. However, its effectiveness is unclear. The aim of this article was to evaluate systematically and critically the effectiveness of SMT for neck pain. Six electronic databases were searched for all relevant randomized clinical trials. Strict inclusion/exclusion criteria had been predefined. Key data were validated and extracted. Methodologic quality was assessed by using the Jadad score.Statistical pooling was anticipated but was deemed not feasible. Four studies met the inclusion/exclusion criteria. Two studies were on single interventions, and 2 included series of SMT treatments, both with a 12-month follow-up. The 2 short-term trials used spinal mobilization as a control intervention. The 2 long-term studies compared SMT with exercise therapy. None of the 4 trials convincingly demonstrated the superiority of SMT over control interventions. In conclusion, the notion that SMT is more effective than conventional exercise treatment in the treatment of neck pain was not supported by rigorous trial data.

Since then, more data have become available. This 2015 Cochrane review assessed the effects of spinal manipulation or mobilisation alone, compared with those of an inactive control or another active treatment, on:

---

[9]Merepeza A. Effects of spinal manipulation versus therapeutic exercise on adults with chronic low back pain: a literature review. *J Can Chiropr Assoc.* 2014;58(4):456–466.

[10]https://www.nice.org.uk/news/press-and-media/exercise-not-acupuncture-for-people-with-low-back-pain-says-nice-in-draft-guidance.

[11]Ernst E. Chiropractic spinal manipulation for neck pain: a systematic review. *J Pain.* 2003;4(8):417–421. https://doi.org/10.1067/s1526-5900(03)00735-1.

- pain,
- function,
- disability,
- patient satisfaction,
- quality of life,
- global perceived effect

in adults experiencing neck pain.[12] The authors included 51 RCTs with a total of 2920 participants. Here are the findings related to SMT:

- Cervical manipulation compared to inactive control: For subacute and chronic neck pain, a single manipulation relieved pain at immediate but not short-term follow-up.
- Cervical manipulation compared to another active treatment: For acute and chronic neck pain, multiple sessions of cervical manipulation produced similar changes in pain, function, quality of life (QoL), global perceived effect (GPE) and patient satisfaction when compared with multiple sessions of cervical mobilisation at immediate-, short- and intermediate-term follow-up. For acute and subacute neck pain, multiple sessions of cervical manipulation were more effective than certain medications in improving pain and function at immediate and long-term follow-up. These findings are consistent for function at intermediate-term follow-up. For chronic CGH, multiple sessions of cervical manipulation may be more effective than massage in improving pain and function at short/intermediate-term follow-up. Multiple sessions of cervical manipulation may be favoured over transcutaneous electrical nerve stimulation (TENS) for pain reduction at short-term follow-up. For acute neck pain, multiple sessions of cervical manipulation may be more effective than thoracic manipulation in improving pain and function at short/intermediate-term follow-up.
- Thoracic manipulation compared to inactive control: Three trials using a single session were assessed at immediate-, short- and intermediate-term follow-up. At short-term follow-up, manipulation improved pain in participants with acute and subacute neck pain and function in participants with acute and chronic neck pain. A funnel plot of these data suggests publication bias. These findings were consistent at intermediate follow-up for pain/function/quality of life.

---

[12]Gross A, Langevin P, Burnie SJ, et al. Manipulation and mobilisation for neck pain contrasted against an inactive control or another active treatment. *Cochrane Database Syst Rev.* 2015;(9):CD004249. Published 2015 Sep 23. https://doi.org/10.1002/14651858.CD004249.pub4.

- Thoracic manipulation versus another active treatment: No studies provided sufficient data for statistical analyses. A single session of thoracic manipulation was comparable with thoracic mobilisation for pain relief at immediate-term follow-up for chronic neck pain.

The authors drew the following conclusions: *Although support can be found for use of thoracic manipulation versus control for neck pain, function and QoL, results for cervical manipulation and mobilisation versus control are few and diverse. Publication bias cannot be ruled out. Research designed to protect against various biases is needed. Findings suggest that manipulation and mobilisation present similar results for every outcome at immediate/short/intermediate-term follow-up. Multiple cervical manipulation sessions may provide better pain relief and functional improvement than certain medications at immediate/intermediate/long-term follow-up. Since the risk of rare but serious adverse events for manipulation exists, further high-quality research focusing on mobilisation and comparing mobilisation or manipulation versus other treatment options is needed to guide clinicians in their optimal treatment choices.*

Thus, the evidence to suggest that chiropractic SMT is effective for neck pain turns out to be weak—too weak to recommend it for routine use. Other options, like neck massages often done by massage therapists, osteopaths or physiotherapists would seem to be be preferable, particularly considering the possibility of harm caused by neck manipulations which we will discuss in Chap. 14.

Box 10.1

Other alternative treatments frequently recommended for back and neck problems; none of them are backed by compelling evidence (for a full evaluation of all these therapies, see[13])

- Acupressure
- Acupuncture
- Alexander technique
- Apitherapy
- Auriculotherapy
- Biopuncture
- Bioresonance
- Bowen technique
- Chondroitin
- Cupping

---

[13] Ernst E. Alternative Medicine: A Critical Assessment of 150 Modalities. Copernicus (2019).

- Dorn therapy
- Energy healing
- Feldenkrais method
- Glucosamine
- Gua sha
- Herbal medicine
- Hot stone massage
- Hypnotherapy
- Magnet therapy
- Massage
- Mind body therapies
- Moxibustion
- Naturopathy
- Neural therapy
- Osteopathy
- Pilates
- Polarity therapy
- Progressive muscle relaxation
- Rolfing
- Shiatsu
- Slapping therapy
- Zero balance
- Traditional Chinese medicine
- Tragerwork
- Tiu na
- yoga

Effectiveness of Spinal Manipulation in Other Problems

Chiropractors correct abnormalities of the intellect as well as those of the body.
— D. D. Palmer

Most consumers find it conceivable that spinal manipulation therapy (SMT) might reduce spinal pain (Chap. 10). However, chiropractors claim to effectively treat many conditions that are not related to the spine, and most people would find this less than plausible. But DD Palmer, the founding father of chiropractic, made very concrete claims about being able to cure a wide range of illnesses (Box 11.1). And most chiropractors find it impossible to doubt their guru.

There is no definitive summary of conditions which today's chiropractors treat, but the list and text quoted below gives you a fairly good impression[1]:

- Headaches, migraines
- Upper, mid, lower back pain & stiffness
- Neck pain & stiffness
- Pinched nerves, muscle spasms
- Leg & arm pain, tingling, numbness, weakness
- Leg & arm joint pain or dysfunction
- Carpal or Tarsal tunnel syndrome.

---

[1] https://chiropractor.com/conditions-chiropractors-treat/.

© Springer Nature Switzerland AG 2020
E. Ernst, *Chiropractic*,
https://doi.org/10.1007/978-3-030-53118-8_11

Many conditions are caused by subluxations (misalignments of spinal bones), which irritate or aggravate the nerves resulting in diseases or conditions. Chiropractic care can provide partial or full relief to many of these conditions:

- Asthma
- Arthritis
- Bursitis
- Chronic Fatigue Syndrome
- Colic
- Ear infection
- Fertility issues
- Frequent colds, cases of flu
- Gastrointestinal syndromes
- Intervertebral disc syndrome
- Loss of equilibrium
- Menstrual disorders
- Multiple Sclerosis
- Sciatica
- Scoliosis
- Tendonitis
- Thoracic outlet syndrome
- Respiratory infection.

Many common injuries can also be effectively treated by chiropractic care. The most common injuries are automobile accidents, where the symptoms can be delayed by weeks or even months after the accident. Auto insurance often covers these injuries. Workplace injuries including falls, lifting injuries, and others are also treatable by chiropractic. Many workplaces offer insurance to cover these types of injuries. Chiropractic works well with sports injuries, and chiropractors can often improve performance. Some common injuries treated by chiropractors include:

- Auto accidents
- Chronic injuries
- Falling injuries
- Lifting injuries
- Sport injuries
- Whiplash
- Work-related injuries.

In this chapter, I will look at some of these claims. As it is hardly possible to be comprehensive—I am not aware of many conditions that chiropractors do <u>not</u> treat—I will focus on such illnesses for which there have been recent studies testing the effectiveness of SMT; others are discussed in Chap. 7.

Migraine

A three-armed, single-blinded, placebo-controlled RCT of 17 months duration included 104 migraineurs.[2] The treatment consisted of SMT (group 1), and the placebo was a sham push manoeuvre of the lateral edge of the scapula and/or the gluteal region (group 2). The third group continued their usual pharmacological management (group 3). The results showed that migraine days were significantly reduced in all three groups. However, the reduction in migraine days was not significantly different between the groups. Migraine duration and headache index were reduced significantly more in the SMT than in group 3 towards the end of follow-up. The authors concluded that *the effect of SMT observed in our study is probably due to a placebo response.*

The findings thus confirm those of our own systematic review suggesting that SMT is not a demonstrably effective therapy for migraine.[3] Three RCTs met our inclusion criteria. Their methodological quality was mostly poor. Two RCTs suggested no effect of SMT in terms of Headache Index or migraine duration and disability compared with drug therapy, SMT plus drug therapy, or mobilization. One RCT showed significant improvements in migraine frequency, intensity, duration and disability associated with migraine compared with detuned interferential therapy. The most rigorous RCT demonstrated no effect of SMT compared with mobilization. We concluded that the *current evidence does not support the use of spinal manipulations for the treatment for migraine headaches.*

[2]Chaibi A, Benth JŠ, Tuchin PJ, Russell MB. Chiropractic spinal manipulative therapy for migraine: a three-armed, single-blinded, placebo, randomized controlled trial. *Eur J Neurol*. 2017;24(1):143–153. http://doi.org/10.1111/ene.13166.

[3]Posadzki P, Ernst E. Spinal manipulations for the treatment of migraine: a systematic review of randomized clinical trials. *Cephalalgia*. 2011;31(8):964–970. http://doi.org/10.1177/033310241140 5226.

Other Forms of Headache

Our 2002 systematic review of SMT for any type of headache included 8 RCTs.[4] Three examined tension-type headaches, three migraine, one 'cervico-genic' headache, and one 'spondylogenic' chronic headache. In two studies, patients receiving SMT showed similar improvements in migraine and tension headaches compared to drug treatment. In the 4 studies employing some form of 'sham' interventions (e.g. laser light therapy), results were less conclusive: two studies showed a benefit for SMT and 2 studies failed to do so. Considerable methodological limitations were observed in most trials, the principal one being inadequate control for placebo effects. Despite claims that SMT is an effective treatment for headache, the evidence did not support such a conclusion. In particular, it was unclear to what extent the observed treatment effects can be explained by any specific effects of SMT or by non-specific factors (e.g. of personal attention, patient expectation). Whether SMT produces any long-term changes was equally uncertain.

The objective of a further systematic review was to assess the effectiveness of spinal manipulations as a treatment option for cervicogenic headaches.[5] Nine RCTs met the inclusion criteria. Their methodological quality was mostly poor. Six RCTs suggested that spinal manipulation is more effective than physical therapy, gentle massage, drug therapy, or no intervention. Three RCTs showed no differences in pain, duration, and frequency of headaches compared to placebo, manipulation, physical therapy, massage, or wait-list controls. Adequate control for placebo effect was achieved in one RCT only, and this trial showed no benefit of spinal manipulations beyond a placebo effect. The majority of RCTs failed to provide details of adverse effects. It was concluded that *the therapeutic value of this approach remains uncertain.*

Otitis Media

Many professional organisations of chiropractic such as the British Chiro-practic Association, the Chiropractic Association of Ireland or the American Chiropractic Association used to state or imply that chiropractic is an effec-

---

[4]Astin JA, Ernst E. The effectiveness of spinal manipulation for the treatment of headache disorders: a systematic review of randomized clinical trials. *Cephalalgia.* 2002;22(8):617–623. http://doi.org/10.1046/j.1468-2982.2002.00423.x.

[5]Posadzki P, Ernst E. Spinal manipulations for cervicogenic headaches: a systematic review of random-ized clinical trials. *Headache.* 2011;51(7):1132–1139. http://doi.org/10.1111/j.1526-4610.2011.01932.x.

tive treatment for ear infections. A recent survey furthermore demonstrated that the majority of UK chiropractors subscribe to this idea.

But is there any evidence that it is true? In an attempt to find out, I conducted a systematic review in 2008.[6] To get included, an article needed to refer to a controlled clinical trial of chiropractic for ear infection (otitis). Case reports, case series and uncontrolled or feasibility studies were excluded. These searches generated 35 hits. After removing duplicates, 27 articles were read in full. None of them met the inclusion criteria. Therefore, the claim that chiropractic SMT is an effective treatment for otitis is not based on evidence from rigorous clinical trials.

Carpal Tunnel Syndrome

A carpal tunnel syndrome is a common condition characterised by pain in the wrist which is particularly strong during night time. It is caused by the tightening of ligaments around the wrist and can be effectively treated with a simple operation releasing the offending ligaments. Yet many chiropractors claim that they can treat the condition without the need of surgery. In 2003, I published a systematic review testing this hypothesis; Here is its abstract[7]:

Background.
Although chiropractic is most commonly used for spinal problems, many chiropractors use manipulations for the treatment of non-spinal conditions. Carpal tunnel syndrome (CTS) has been identified as one such condition. This systematic review evaluates the evidence for or against the effectiveness of chiropractic as a treatment for CTS.

Methods.
Eight electronic databases were searched from inception until November 2008. Reference lists of retrieved articles were hand-searched. Chiropractic associations were contacted in order to identify further non-published studies. No language restrictions were applied.

Results.
Of 26 potentially relevant studies, only one trial of chiropractic for CTS met all the inclusion criteria. The trial was of poor quality and reported no significant differences between the groups on any outcome measure.

---

[6]Ernst E. Re: Chiropractic for otitis?. *Int J Clin Pract*. 2009;63(9):1393. http://doi.org/10.1111/j. 1742-1241.2009.02097.x.
[7]Hunt KJ et al.: Hand Therapy 2009;14:89–94.

However, our re-analyses indicated a significant difference in favour of the control treatment (non-steroidal anti-inflammatory drugs [NSAIDs] use). Adverse effects were noted in both groups.

Conclusions.
There is insufficient evidence to suggest that chiropractic is effective for the treatment of CTS. Therapy should continue to focus on the use of NSAIDs, corticosteroid injection, splinting and surgical release of the median nerve. Further research into the utility of chiropractic for CTS is required.

Fibromyalgia

Many patients use chiropractic as a treatment of fibromyalgia, and many chiropractors seem to be convinced that it is effective for that condition. A systematic review evaluated the effectiveness of chiropractic care for fibromyalgia.[8] Three RCTs met the inclusion criteria. Their methodological quality was poor. Collectively, they generated no sound evidence to suggest that chiropractic care is effective for fibromyalgia. It was concluded that *there is insufficient evidence to conclude that chiropractic is an effective treatment for fibromyalgia.*

Treatment of Shoulder Problems

This 2017 review investigated SMT as a treatment for shoulder problems.[9] The intervention included SMT, as well as manipulative therapy directed to the shoulder and/or the regions of the cervical or thoracic spine. Six trials with patients suffering from subacromial impingement syndrome met inclusion criteria. Four studies were RCTs, and two were uncontrolled observational studies. Three of 4 RCTs compared a thrust manipulation to sham SMT. No pain reduction was found between SMT and sham treatment. The authors concluded that *there is limited evidence to support or refute thrust manipulation as a solitary treatment for shoulder pain or disability associated with subacromial impingement syndrome. Studies consistently reported a reduction in pain and*

---

[8] Ernst E. Chiropractic treatment for fibromyalgia: a systematic review. *Clin Rheumatol.* 2009;28(10):1175–1178. http://doi.org/10.1007/s10067-009-1217-9.

[9] Minkalis AL, Vining RD, Long CR, Hawk C, de Luca K. A systematic review of thrust manipulation for non-surgical shoulder conditions. *Chiropr Man Therap.* 2017;25:1. Published 2017 Jan 4. http://doi.org/10.1186/s12998-016-0133-8.

improvement in disability following thrust manipulation. In RCTs, active treatments were comparable to shams suggesting that addressing impingement issues by manipulation alone may not be effective. Thrust manipulative therapy appears not to be harmful, but AE reporting was not robust. Higher-quality studies with safety data, longer treatment periods and follow-up outcomes are needed to develop a stronger evidence-based foundation for thrust manipulation as a treatment for shoulder conditions. These conclusions require some clarifications:

- The review included uncontrolled studies which tell us little about the effectiveness of the treatments in question (Chap. 9).
- Pain reductions were found within groups but not between SMT and sham. This implies that SMT is a placebo therapy.
- Claiming that there was limited evidence to support or refute SMT as a solitary treatment for shoulder pain is not justified by the data. In fact, there is no good positive evidence.
- Stating 'thrust manipulative therapy appears not to be harmful, but AE reporting was not robust' is misleading. Even if there had been adequate reporting of side-effects, and even if this had not disclosed any problems, the safety of SMT (or any other intervention) cannot be judged on the basis of such a small sample (Chap. 14).

Osteoarthritis

A 2013 study aimed at comparing the effectiveness of three different interventions, all administered for 6 weeks:

1. a patient education (PEP) programme,
2. a patient education (PEP) programme plus chiropractic SMT,
3. a minimal control intervention (MCI).[10]

A total of 118 patients with hip osteoarthritis (OA) were included. The PEP was taught by a physiotherapist in 5 sessions. The SMT was delivered by a chiropractor in 12 sessions, and the MCI included a home stretching programme. The primary outcome measure was the self-reported pain severity following the intervention period and after one year. The

---

[10]Poulsen E, Hartvigsen J, Christensen HW, Roos EM, Vach W, Overgaard S. Patient education with or without manual therapy compared to a control group in patients with osteoarthritis of the hip. A proof-of-principle three-arm parallel group randomized clinical trial. *Osteoarthritis Cartilage.* 2013;21(10):1494–1503. http://doi.org/10.1016/j.joca.2013.06.009.

primary analyses included 111 patients. In the PEP + SMT group, a statistically and clinically significant reduction in pain severity of 1.9 points was noted compared to the MCI control group. No difference was found between the PEP and the MCI groups. At 12 months, the difference favouring PEP + SMT was maintained. The authors concluded that *for primary care patients with osteoarthritis of the hip, a combined intervention of manual therapy and patient education was more effective than a minimal control intervention. Patient education alone was not superior to the minimal control intervention.*

These findings can be interpreted in at least two very different ways. One explanation would be that chiropractic SMT is effective. Yet, an alternative explanation seems more plausible: the added time, attention and encouragement provided by the chiropractor (who must have been aware of what was at stake and hence highly motivated) was the effective element in the SMT intervention, while the SMT per se made little or no difference. The PEP + SMT group had no less than 12 sessions with the chiropractor. It seems likely that this additional care, compassion, empathy, time, encouragement etc. was a crucial factor in making these patients feel better and in convincing them to adhere more closely to the instructions of the PEP.

Our systematic review evaluated the evidence on the efficacy and effectiveness of all practitioner-based complementary therapies (including chiropractic) for patients with OA. In all, 16 eligible trials were identified covering 12 therapies. Overall, there was no good evidence of the effectiveness of any of the therapies in relation to pain or global health improvement/quality of life because most therapies only had a single randomized controlled trial. We concluded that *there is not sufficient evidence to recommend any of the practitioner-based complementary therapies considered here for the management of OA, but neither is there sufficient evidence to conclude that they are not effective or efficacious.*[11]

Sport Injuries

Many chiropractors claim that they can effectively treat a range of sports injuries, and numerous sport clubs employ chiropractors to look after their members. Our systematic review of 2011 assessed the effectiveness of chiro-

---

[11]Macfarlane GJ, Paudyal P, Doherty M, et al. A systematic review of evidence for the effectiveness of practitioner-based complementary and alternative therapies in the management of rheumatic diseases: osteoarthritis. *Rheumatology (Oxford)*. 2012;51(12):2224–2233. http://doi.org/10.1093/rheumatology/kes200.

practic interventions for the treatment and/or prevention of sports injuries.[12] Four RCTs and two non-randomised controlled clinical trials met the inclusion criteria. The methodological quality of all the included studies was poor. Three studies suggested that chiropractic was an effective treatment for sports injuries. Two RCTs indicated that there was no difference between chiropractic and control groups in the treatment of sports injuries. One RCT suggested that chiropractic was not effective for the prevention of sports injuries. We concluded that *few rigorous trials have tested the effectiveness of chiropractic manipulation for the treatment and/or prevention of sports injuries. Thus, the therapeutic value of this approach for athletes remains uncertain.*

Any Non-spinal Pain

According to chiropractic theory, SMT should effectively treat any type of pain (Chaps. 4 and 5). It is therefore hardly surprising that many chiropractors make such claims and treat patients suffering from generalised pain. But is the claim true? In 2003, I conducted a systematic review to find out; here is the abstract[13]:

Aims:
Chiropractic manipulation is mostly used for spinal problems but, in an increasing number of cases, also for non-spinal conditions. This systematic review is aimed at critically evaluating the evidence for or against the effectiveness of this approach.

Methods:
Five electronic databases were searched for all randomised clinical trials of chiropractic manipulation as a treatment of non-spinal pain. They were evaluated according to standardised criteria.

Results:
Eight such studies were identified. They related to the following conditions: fibromyalgia, carpal tunnel syndrome, infantile colic, otitis media, dysmenorrhoea and chronic pelvic pain. Their methodological quality ranged from mostly poor to excellent. Their findings do not demonstrate that chiropractic manipulation is an effective therapy for any of these conditions.

---

[12] Posadzki P, Ernst E: Focus on Alternative and Complementary Therapies 2012;17:9–14.

[13] Ernst E. Chiropractic manipulation for non-spinal pain—a systematic review. *N Z Med J.* 2003;116(1179):U539. Published 2003 Aug 8.

Conclusions:
Only very few randomised clinical trials of chiropractic manipulation as a treatment of non-spinal conditions exist. The claim that this approach is effective for such conditions is not based on data from rigorous clinical trials.

Opioid Over-Use

Opioids are potent pain-killers but they can cause major problems, not least dependence. Therefore, much effort is directed towards preventing opioid-related problems. Chiropractors were quick to claim that their treatments might help to avoid them. This systematic review determined whether there is an association between chiropractic use and the frequency of opioid prescriptions.[14] The prevalence of chiropractic care among patients with spinal pain varied between 11 and 51%. Chiropractic users had 64% lower odds of receiving an opioid prescription than nonusers. The authors concluded that *this review demonstrated an inverse association between chiropractic use and opioid receipt among patients with spinal pain. Further research is warranted to assess this association and the implications it may have for case management strategies to decrease opioid use.*

But what do such findings tell us? The important thing to remember here is CORRELATION IS NOT CAUSATION!

- People who drive a VW car are less likely to buy a Mercedes.
- People who have ordered fish in a restaurant are unlikely to also order a steak.
- People who use physiotherapy for back pain will probably use less opioids than those who don't consult physios.
- People who treat their back pain with massage therapy are less likely to also use opioids.

An interesting finding from this analysis is that 31–66% of patients using chiropractic for their neck/back pain also took opioids. This seems to confirm that chiropractic SMT is not nearly as effective as we are often led to believe (Chap. 10).

---

[14]Corcoran KL, Bastian LA, Gunderson CG, Steffens C, Brackett A, Lisi AJ. Association Between Chiropractic Use and Opioid Receipt Among Patients with Spinal Pain: A Systematic Review and Meta-analysis. *Pain Med*. 2020;21(2):e139–e145. http://doi.org/10.1093/pm/pnz219.

Another US study with low back pain (LBP) patients examined the association of initial provider treatment with opioid use.[15] The 216 504 patients were aged 18 years or older and had been diagnosed with new-onset LBP and were opioid-naïve. The primary independent variable was the type of initial healthcare provider including physicians and conservative therapists (physical therapists, chiropractors, acupuncturists). The main outcome measures were short-term opioid use (within 30 days of the index visit) following new LBP visit and long-term opioid use (starting within 60 days of the index date and either 120 or more days' supply of opioids over 12 months, or 90 days or more supply of opioids and 10 or more opioid prescriptions over 12 months). Short-term use of opioids was 22%. Patients who received initial treatment from chiropractors or physical therapists had decreased odds of short-term and long-term opioid use compared with those who received initial treatment from primary care physicians. The authors concluded *that initial visits to chiropractors or physical therapists is associated with substantially decreased early and long-term use of opioids. Incentivising use of conservative therapists may be a strategy to reduce risks of early and long-term opioid use.*

As in the previously discussed paper, the nature of the association remains unclear. Is it correlation or causation? Concluding that initial visits to chiropractors or physical therapists is associated with substantially decreased early and long-term use of opioids implies a causal relationship. But causality remains unproven. Likewise, it is odd to claim that incentivising the use of chiropractors or physiotherapists may reduce the risk of opioid use. The only thing that reduces opioid use is a reduction of opioid prescribing. The way to achieve this is to teach and train doctors adequately and remind them that opioids are not normally for long-term use.

Hypertension

A survey suggested that 42% of UK chiropractors believe they can effectively treat hypertension.[16] As hypertension is a common and serious risk factor for cardiovascular events, it was important to conduct a systematic review to check whether this belief is based on sound evidence.[17] The review included 4

[15] Kazis LE, Ameli O, Rothendler J, et al. Observational retrospective study of the association of initial healthcare provider for new-onset low back pain with early and long-term opioid use [published correction appears in BMJ Open. 2020 Jan 10;10(1):e028633corr1]. *BMJ Open.* 2019;9(9):e028633. Published 2019 Sep 20. http://doi.org/10.1136/bmjopen-2018-028633.

[16] Pollentier A, Langworthy JM: The scope of chiropractic practice. Clin Chiropractic 2007;10:147–155.

[17] Ernst E: Chiropractic manipulation as a treatment of hypertension? Perfusion 2008;21:188–190.

RCTs that had tested the hypothesis. Their methodological rigour was mostly poor. The totality of the evidence did not show that chiropractic lowers blood pressure. It was concluded that, *until evidence to the contrary emerges, chiropractic SMT cannot be considered an effective treatment for hypertension.*

Gastrointestinal Problems

Many chiropractors believe that chiropractic treatments are effective for gastrointestinal disorders. The aim of another of my own systematic reviews was to critically evaluate the evidence from controlled clinical trials supporting or not supporting this notion.[18] Prospective, controlled, clinical trials of any type of chiropractic treatment for any type of gastrointestinal problem, except infant colic (this chapter), were included. Only two trials were found. One was a mere pilot study, but the other was a proper trial. Its authors had reached a positive conclusion; however, their study had serious methodological flaws which made it uninterpretable. My conclusion therefore was that *there is no supportive evidence that chiropractic is an effective treatment for gastrointestinal disorders.*

The evidence discussed in this chapter thus fails to provide sound evidence that chiropractic is an effective option for the management of non-spinal conditions. In the following chapter, we will see whether this applies only to adult patients or whether it is also true for children.

Box 11.1

Examples of conditions which DD Palmer claimed to be able to cure with spinal adjustments[19]:

- Apoplexy
- Asthma
- Bronchitis
- Bunions
- Chickenpox
- Cholera
- Diabetes
- Dyspepsia

---

[18] Ernst E. Chiropractic treatment for gastrointestinal problems: a systematic review of clinical trials. *Can J Gastroenterol*. 2011;25(1):39–40. http://doi.org/10.1155/2011/910469.

[19] Palmer DD. *Text-Book of the Science, Art and Philosophy of Chiropractic*. Rev. edn. Echo Point Books & Media (2019).

- Endocarditis
- Epilepsy
- Gout
- Hydrocephalus
- Insanity
- Malaria
- Meningitis
- Mumps
- Neuropathy
- Paralysis
- Peritonitis
- Pleuritis
- Pneumonia
- Syphilis
- Tuberculosis
- Worms

12

Chiropractic for Children

Many a child has been injured at birth by a vertebral displacement which caused an impingement upon one or more of the spinal nerves.
—DD Palmer

As chiropractors are lobbying to become accepted as primary care physicians (Chap. 7), it is hardly surprising that most also claim to be able to treat young children and babies and that 39% of US chiropractors engage in paediatric care.[1] Elsewhere, the situation is similar; a survey of UK chiropractors, for instance, suggested that 10–58% of the respondents considered conditions like infantile colic, childhood asthma, enuresis, otitis, epilepsy, attention deficit hyperactivity disorder (ADHD) or cerebral palsy to be "effectively treatable by chiropractic methods".[2] One website[3] went further by listing the 'Top 5 Reasons To Take Your Child To A Chiropractor':

1. To promote proper growth and development and to limit health challenges such as issues with nursing and reflux.
2. To allow a child's nervous system and spine to grow optimally and without interference. To reduce ear infections, bed wetting, asthma and allergies.

---

[1] Document available at https://mynbce.org/wp-content/uploads/2020/02/Executive-Summary-Practice-Analysis-of-Chiropractic-2020.pdf.

[2] Pollentier A, Langworthy JM. The scope of chiropractic practice: a survey of chiropractors in the UK. Clin Chiropractic 2007;10:147–55.

[3] https://www.onelovechiropractic.com/top-5-reasons-to-take-your-child-to-the-chiropractor/.

© Springer Nature Switzerland AG 2020
E. Ernst, *Chiropractic*,
https://doi.org/10.1007/978-3-030-53118-8_12

3. To improve a child's immune system and digestive health and to limit colic and constipation.
4. To encourage nerve and brain development (neural plasticity). To promote proper perception and awareness preventing labels such as ADHD, Sensory Integration Disorder and other neurodevelopmental disorders.
5. To be proactive about your child's overall health and wellbeing which will limit and support them through colds and infections and allow them to meet their optimal potential.

But how solid is the evidence for these claims? Seasoned chiropractor Samuel Homola is sceptical[4]:

Chiropractors commonly treat children for a variety of ailments by manipulating the spine to correct a 'vertebral subluxation' or a 'vertebral subluxation complex' alleged to be a cause of disease. Such treatment might begin soon after a child is born. Both major American chiropractic associations - the International Chiropractic Association and the American Chiropractic Association - support chiropractic care for children, which includes subluxation correction as a treatment or preventive measure. I do not know of any credible evidence to support chiropractic subluxation theory. Any attempt to manipulate the immature, cartilaginous spine of a neonate or a small child to correct a putative chiropractic subluxation should be regarded as dangerous and unnecessary. Referral of a child to a chiropractor for such treatment should not be considered lest a bad outcome harms the child or leads to a charge of negligence or malpractice.

Sadly, Homola's views are not shared by the majority of his colleagues. A recent survey looked at the reports by mothers of their infants' condition before and after chiropractic care.[5] A total of 16 UK chiropractic clinics participated, 2001 mothers filled the questionnaires, and 1092 completed the follow-up forms. They reported improvements across all aspects of infant behaviour, including:

- feeding problems,
- sleep issues,
- excessive crying,

---

[4]Homola S. Pediatric Chiropractic Care: The Subluxation Question and Referral Risk. *Bioethics*. 2016;30(2):63–68. https://doi.org/10.1111/bioe.12225.
[5]Miller JE, Hanson HA, Hiew M, Lo Tiap Kwong DS, Mok Z, Tee YH. Maternal Report of Outcomes of Chiropractic Care for Infants. *J Manipulative Physiol Ther*. 2019;42(3):167–176. https://doi.org/10.1016/j.jmpt.2018.10.005.

- problems with supine sleep position,
- infant pain,
- restricted cervical range of motion,
- time performing prone positioning.

Maternal ratings of depression, anxiety, and satisfaction with motherhood also demonstrated statistically significant improvement. In total, 82% reported definite improvement of their infants on a global impression of change scale. In addition, 95% felt that the care was cost-effective, and 90.9% rated their satisfaction 8 or higher on an 11-point scale. The authors concluded that *mothers reported that chiropractic care for their infants was effective, safe, and cost-effective.*

As much as such data are liked by chiropractors, they are less than convincing. Here are some of the most obvious caveats:

- The fact that mothers reported positive outcomes is hardly surprising. After all, they chose to consult a chiropractor and paid for the treatments. Install yourself in a McDonald's, ask customers whether they like hamburgers, and I am sure you will get a similarly meaninglessly positive result.
- Cost-effectiveness cannot be measured by asking people questions like 'was it worth it?'.
- A survey of this nature does not allow any conclusions about cause and effect (Chap. 9).
- The non-validated nature of the outcome measures means that the reliability of the findings is questionable.

In 2009, I surveyed 5 professional organisations of chiropractic and looked for statements about chiropractic care for children.[6] These are the statements I found:

- THE BRITISH CHIROPRACTIC ASSOCIATION: "As children grow, chiropractic can help not only the strains caused by the rough and tumble of life but also with some of the problems that children can suffer in their first years: Colic—sleeping and feeding problems—frequent ear infections—asthma—prolonged crying."
- THE CHIROPRACTIC ASSOCIATION OF IRELAND: "Even today's 'natural' childbirth methods can affect an infant's spine. Preliminary studies suggested that colic, unusual crying, poor appetite, ear infections or erratic

---

[6]Ernst E. Chiropractic for children. *Pediatrics.* 2008;122(5):1161. https://doi.org/10.1542/peds.2008-2620.

sleeping habits can be signs of spinal distress. Paediatric adjustments are gentle. Knowing exactly where to adjust, the doctor applies no more pressure than you'd use to test the ripeness of a tomato."

- THE McTIMONEY CHIROPRACTIC ASSOCIATION: "Birth is probably one of the toughest events we undergo as humans. A baby's head has to squeeze through a small birth canal to be born. In doing so the baby's head in particular will absorb much of the shock, and the soft bones will yield slightly allowing it to travel down the birth canal. This is called 'moulding.' After birth the baby's head will gradually revert to a more normal shape. However, if this 'unmoulding' does not take place completely, the baby can be left in some discomfort which they are unable to communicate. Most babies cope extremely well with the process and emerge contented, happy, able to feed, sleep, and grow normally. However, for some, the recovery can take longer, especially those who had a particularly difficult entry into the world and these babies may show some, all, or a combination of the following signs:

  - Irritability, fractiousness
  - Feeding problems
  - Continuous crying
  - Sleeps little, difficult to settle
  - Colic, sickness and wind

  All of these could indicate that there is a misalignment in the baby's skeletal system and that the baby is uncomfortable as a result. These misalignments could be causing discomfort both when lying down and when lifted, hence many parents report that whatever they do, whether they lift their baby up or lie him down, it seems to make no difference, and the crying continues. Feeding problems may indicate that there is a problem with the nerves at the base of the skull and that the digestive system is compromised, or more simply the baby may be uncomfortable sucking due to mechanical stresses on its skeletal structure.

- THE NEWZEALAND CHIROPRACTOR ASSOCIATION: "Parents and teachers often report children dramatically improve their attention span with Chiropractic care. Others report Chiropractic care as being the main reason for improved academic performance."
- THE SCOTTISH CHIROPRACTOR SASSOCIATION: "Chiropractors are able to examine and evaluate a child's spine to determine if they can help problems such as colic, asthma, bedwetting, eczema and sleeping difficulties. Chiropractors advise that a child's spine be checked

for subluxations and postural distortions before any symptoms are even present."

I was unable to find convincing evidence for any of the above-named conditions or claims.[7] More recently, a systematic review shed more light on the issue by evaluating the use of manual therapy for clinical conditions in the paediatric population, assessing the methodological quality of the studies found, and synthesizing findings based on health condition.[8] A total of 50 studies (32 RCTs and 18 observational studies) met the authors' inclusion criteria. Only 18 studies were judged to be of high quality. Conditions evaluated were:

- attention deficit hyperactivity disorder (ADHD),
- autism,
- asthma,
- cerebral palsy,
- clubfoot,
- constipation,
- cranial asymmetry,
- cuboid syndrome,
- headache,
- infantile colic,
- low back pain,
- obstructive apnoea,
- otitis media,
- paediatric dysfunctional voiding,
- paediatric nocturnal enuresis,
- postural asymmetry,
- preterm infants,
- pulled elbow,
- suboptimal infant breastfeeding,
- scoliosis,
- suboptimal infant breastfeeding,
- temporomandibular dysfunction,

---

[7]Ernst E. Chiropractic for paediatric conditions: substantial evidence?. *BMJ*. 2009;339:b2766. Published 2009 Jul 9. https://doi.org/10.1136/bmj.b2766.

[8]Parnell Prevost C, Gleberzon B, Carleo B, Anderson K, Cark M, Pohlman KA. Manual therapy for the pediatric population: a systematic review. *BMC Complement Altern Med*. 2019;19(1):60. Published 2019 Mar 13. https://doi.org/10.1186/s12906-019-2447-2.

- torticollis,
- upper cervical dysfunction.

Musculoskeletal conditions, including low back pain and headache, were evaluated in seven studies. Only 20 studies reported adverse events. The authors concluded that *fifty studies investigated the clinical effects of manual therapies for a wide variety of pediatric conditions. Moderate-positive overall assessment was found for 3 conditions: low back pain, pulled elbow, and premature infants. Inconclusive unfavorable outcomes were found for 2 conditions: scoliosis (OMT) and torticollis (MT). All other condition's overall assessments were either inconclusive favorable or unclear. Adverse events were uncommonly reported. More robust clinical trials in this area of healthcare are needed.*

There are several things that are remarkable about this review:

- The list of indications for which studies have been published demonstrates that chiropractors regard their approach as a panacea.
- A systematic review evaluating the effectiveness of a therapy that includes observational studies without a control group is nonsensical (Chap. 9).
- Many of the RCTs included in the review are meaningless; for instance, if a trial compares the effectiveness of two different manual therapies neither of which has been shown to work, it cannot generate a meaningful result.
- The majority of trialists failed to report adverse effects. This is a violation of medical ethics and casts doubt on the validity of the study. One could even argue that such trials are best discarded.
- Only three conditions are, according to the review authors, based on evidence. This is hardly enough to justify the existence of an entire speciality of paediatric chiropractors.

A closer look at these three allegedly evidence-based paediatric conditions seems justified.

1. **Low back pain**: the verdict 'moderate positive' is based on two randomised clinical trials (RCTs) and two observational studies. The latter are irrelevant for evaluating the effectiveness of a therapy (Chap. 9). One of the two RCTs should have been excluded because the age of the patients exceeded the age range named by the authors as an inclusion criterion. This leaves us with one single 'medium quality' RCT that included a mere 35 patients. It would be more than foolish to base a positive verdict on such evidence.

2. **Pulled elbow**: the positive verdict is based on one single RCT that compared two different approaches of unknown value. It is not justified to base a positive verdict on such evidence.

3. **Preterm**: for this condition, there were 4 RCTs; one was a pilot study of craniosacral therapy (Chap. 8) with a study design that does not allow for control for placebo effects. The other three RCTs were all from the same Italian research group; their findings have never been independently replicated. It would be irresponsible to base a positive verdict on such flimsy evidence.

Another systematic review evaluated the evidence for effectiveness and harms of specific SMT techniques for infants, children and adolescents.[9] In total, 26 studies were included. Infants and children/adolescents were treated for various (non-)musculoskeletal indications, hypothesized to be related to spinal joint dysfunction. The results showed that:

- Due to very low quality evidence, it is uncertain whether gentle, low-velocity mobilizations reduce complaints in infants with colic or torticollis, and whether high-velocity, low-amplitude manipulations reduce complaints in children/adolescents with autism, asthma, nocturnal enuresis, headache or idiopathic scoliosis.
- Five case reports described severe harm after HVLA manipulations in 4 infants and one child. Mild, transient harm was reported after gentle spinal mobilizations in infants and children and could be interpreted as a side effect of the treatment.

The authors concluded that *due to very low quality of the evidence, the effectiveness of gentle, low-velocity mobilizations in infants and HVLA manipulations in children and/or adolescents is uncertain. Assessments of intermediate outcomes are lacking in current pediatric SMT research. Therefore, the relationship between specific treatment and its effect on the hypothesized spinal dysfunction remains unclear. Gentle, low-velocity spinal mobilizations seem to be a safe treatment technique. Although scarcely reported, HVLA manipulations in infants and young children could lead to severe harms. Severe harms were likely to be associated with unexamined or missed underlying medical pathology.*

[9]Driehuis F, Hoogeboom TJ, Nijhuis-van der Sanden MWG, de Bie RA, Staal JB. Spinal manual therapy in infants, children and adolescents: A systematic review and meta-analysis on treatment indication, technique and outcomes. *PLoS One.* 2019;14(6):e0218940. Published 2019 Jun 25. https://doi.org/10.1371/journal.pone.0218940.

The question whether SMT is an effective treatment for infant colic has particular significance, since it has attracted much attention in recent years. The reason for this is, of course, that a few years ago Simon Singh had disclosed that the British Chiropractic Association (BCA) was promoting chiropractic treatment for this and 6 other childhood conditions on their website. He famously wrote "*they* (the BCA) *happily promote bogus treatments*" and was subsequently sued for libel by the BCA. Eventually, the BCA lost the libel action as well as lots of money and reputation.

My own review of 2009 concluded that the infant colic claim *is not based on convincing data from rigorous clinical trials.*[10] In 2012, a Cochrane review of manipulative therapies for infant colic was published.[11] Here are its conclusions:

> The studies involved too few participants and were of insufficient quality to draw confident conclusions about the usefulness and safety of manipulative therapies. Although five of the six trials suggested crying is reduced by treatment with manipulative therapies, there was no evidence of manipulative therapies improving infant colic when we only included studies where the parents did not know if their child had received the treatment or not. No adverse effects were found, but they were only evaluated in one of the six studies.

In other words, there is no reliable evidence to suggest that SMT is effective for infant colic or any other paediatric condition.

Potential for Harm

As chiropractic has not been shown to be effective for any paediatric condition, the risks of SMT (Chaps. 14 and 15) are particularly relevant for children who are vulnerable and too young to give informed consent to SMT.

The purpose of this review was to summarise the case reports of adverse effects (AEs) in infants and children treated by chiropractors or other manual therapists.[12] Thirty-one articles met the inclusion criteria. A total of 12

[10] Ernst E. Chiropractic spinal manipulation for infant colic: a systematic review of randomised clinical trials. *Int J Clin Pract.* 2009;63(9):1351–1353. https://doi.org/10.1111/j.1742-1241.2009.02133.x.

[11] Dobson D, Lucassen PL, Miller JJ, Vlieger AM, Prescott P, Lewith G. Manipulative therapies for infantile colic. *Cochrane Database Syst Rev.* 2012;12:CD004796. Published 2012 Dec 12. https://doi.org/10.1002/14651858.CD004796.pub2.

[12] Todd AJ, Carroll MT, Robinson A, Mitchell EKL. Adverse Events Due to Chiropractic and Other Manual Therapies for Infants and Children: A Review of the Literature. *J Manipulative Physiol Ther.* 2015;38(9):699–712. https://doi.org/10.1016/j.jmpt.2014.09.008.

articles reporting 15 serious AEs, including three fatalities, were found. High-velocity, extension, and rotational SMT was reported in most cases. The authors concluded *that published cases of serious adverse events in infants and children receiving chiropractic, osteopathic, physiotherapy, or manual medical therapy are rare. The 3 deaths that have been reported were associated with various manual therapists; however, no deaths associated with chiropractic care were found in the literature to date. Because underlying preexisting pathology was associated in a majority of reported cases, performing a thorough history and examination to exclude anatomical or neurologic anomalies before applying any manual therapy may further reduce adverse events across all manual therapy professions.*

In this context, the statement from the 'Spanish Association of Paediatric Medicines Committee' is of particular value and importance[13]:

Currently, there are some therapies that are being practiced without adjusting to the available scientific evidence. The terminology is confusing, encompassing terms such as "alternative medicine", "natural medicine", "complementary medicine", "pseudoscience" or "pseudo-therapies". The Medicines Committee of the Spanish Association of Paediatrics considers that no health professional should recommend treatments not supported by scientific evidence. Also, diagnostic and therapeutic actions should be always based on protocols and clinical practice guidelines. Health authorities and judicial system should regulate and regularize the use of alternative medicines in children, warning parents and prescribers of possible sanctions in those cases in which the clinical evolution is not satisfactory, as well responsibilities are required for the practice of traditional medicine, for health professionals who act without complying with the "lex artis ad hoc", and for the parents who do not fulfill their duties of custody and protection. In addition, it considers that, as already has happened, Professional Associations should also sanction, or at least reprobate or correct, those health professionals who, under a scientific recognition obtained by a university degree, promote the use of therapies far from the scientific method and current evidence, especially in those cases in which it is recommended to replace conventional treatment with pseudo-therapy, and in any case if said substitution leads to a clinical worsening that could have been avoided.

[13]Piñeiro Pérez R, Núñez Cuadros E, Rodríguez Marrodan B, et al. Posicionamiento del Comité de Medicamentos de la Asociación Española de Pediatría en relación con el uso de medicinas alternativas y seudociencias en niños [Position Statement from the Spanish Association of Paediatric Medicines Committee Concerning the Use of Alternative Medicine and Pseudo-Science in Children]. *An Pediatr (Barc)*. 2019;91(4):272.e1-272.e5. https://doi.org/10.1016/j.anpedi.2019.04.013.

In 2019, the Council of Australian Governments (COAG) Health Council (CHC) noted concerns about SMT on children performed by chiropractors and agreed that there was a need to consider whether public safety was at risk (Box 12.1). Thus they commissioned an independent review.[14] It concluded that *it is clear that spinal manipulation in children is not wholly without risk. Any risk associated with care, no matter how uncommon or minor, must be considered in light of any potential or likely benefits. This is particularly important in younger children, especially those under the age of 2 years in whom minor adverse events may be more common.*

After careful consideration of this collective evidence, it seems clear that the best evidence available to date fails to demonstrate clinically relevant benefits of chiropractic SMT for paediatric patients. Further evidence suggests that chiropractors can cause serious harm to children. In the interest of vulnerable children, we should not be manipulated by misleading statements to the contrary[15] but recommend avoiding chiropractic care for children.

Box 12.1 Excerpts from the Report by the Council of Australian Governments

Based on this review of effectiveness, spinal manipulation of children cannot be recommended for:

- headache
- asthma
- otitis media
- cerebral palsy
- hyperactivity disorders
- torticollis

The Board expects practitioners to:

- discuss their proposed management plan with the patient's parent and/or guardian
- inform the parent and/or guardian about the quality of the acceptable evidence and explain the basis for the proposed treatment
- provide patients (or parent and/or guardian) with information about the risks and benefits of the proposed treatment and the risks of receiving no treatment
- understand that children have significant anatomical, physiological, developmental and psychological differences and needs from adults and that

---

[14] Document available at https://www.bettersafercare.vic.gov.au/sites/default/files/2019-10/20191024-Final%20Chiropractic%20Spinal%20Manipulation.pdf.

[15] Ernst E. Chiropractic manipulation, with a deliberate "double entendre". *Arch Dis Child.* 2009;94(6):411. https://doi.org/10.1136/adc.2009.158170.

their healthcare management requires specific skills and expertise; including informed consent, examination, diagnosis, referral of 'red flags' and contraindications to care

- modify all care and treatment (including technique and force) to suit the age, presentation and development of the patient
- promptly refer patients to the care of other registered health practitioners when they have conditions or symptoms outside a chiropractor's scope of practice, for example 'red flags', and
- communicate effectively with other health practitioners involved with the care of the patient such as the patient's general practitioner or paediatrician.

13

Disease Prevention

You don't just go to the dentist when you have a cavity, do you? Regular chiropractic care is maintenance for your spine.
—statement on Twitter by a chiropractor

Preventing disease is obviously better than curing it. Chiropractors therefore claim that they put a particular focus on disease prevention (Box 13.1). The US 'Practice Analysis of Chiropractic 2020' showed that 65% of US chiropractors focus on wellness and maintenance care.[1] A survey published in a leading chiropractic journal agrees; here is its abstract[2]:

Objective: To investigate the primary care, health promotion activities associated with what has historically been called "maintenance care" (MC) as used in the practice of chiropractic in the United States. This includes issues such as investigating the purpose of MC, what conditions and patient populations it best serves, how frequently it is required, what therapeutic interventions constitute MC, how often it is recommended, and what percent of patient visits are for prevention and health promotion services. It also investigates the economic impact of these services.

---

[1] Document available at https://mynbce.org/wp-content/uploads/2020/02/Executive-Summary-Practice-Analysis-of-Chiropractic-2020.pdf.

[2] Rupert RL. A survey of practice patterns and the health promotion and prevention attitudes of US chiropractors. Maintenance care: part I. *J Manipulative Physiol Ther*. 2000;23(1):1-9. https://doi.org/10.1016/s0161-4754(00)90107-6.

© Springer Nature Switzerland AG 2020
E. Ernst, *Chiropractic*,
https://doi.org/10.1007/978-3-030-53118-8_13

Design: Postal survey of a randomized sample of practicing US chiropractors. The questionnaire was structured with a 5-point ordinal Likert scale (28 questions) and brief fill-in questionnaire (12 questions). The 40-question survey was mailed to 1500 chiropractors selected at random from a pool of chiropractors with active practices in the United States. The National Directory of Chiropractic database was the source of actively practicing chiropractors from which doctor selection was made. The sample was derived by using the last numbers composing the zip codes assigned by the US Postal Service. This sampling method assured potential inclusion of chiropractors from all 50 states, from rural areas and large cities, and assured a sample weighting based on population density that might not have been afforded by a simple random sample.

Results: Six hundred and fifty-eight (44%) of the questionnaires were completed and returned. US chiropractors agreed or strongly agreed that the purpose of MC was to optimize health (90%), prevent conditions from developing (88%), provide palliative care (86%), and minimize recurrence or exacerbations (95%). MC was viewed as helpful in preventing both musculoskeletal and visceral health problems. There was strong agreement that the therapeutic composition of MC placed virtually equal weight on exercise (96%) and adjustments/manipulation (97%) and that other interventions, including dietary recommendations (93%) and patient education about lifestyle changes (84%), shared a high level of importance. Seventy-nine percent of chiropractic patients have MC recommended to them and nearly half of those (34%) comply. The average number of recommended MC visits was 14.4 visits per year, and the total revenue represents an estimated 23% of practice income.

Conclusions: Despite educational, philosophic, and political differences, US chiropractors come to a consensus about the purpose and composition of MC. Not withstanding the absence of scientific support, they believe that it is of value to all age groups and a variety of conditions from stress to musculoskeletal and visceral conditions. This strong belief in the preventive and health promotion value of MC motivates them to recommend this care to most patients. This, in turn, results in a high level of preventive services and income averaging an estimated $50,000 per chiropractic practice in 1994. The data suggest that the amount of services and income generated by preventive and health-promoting services may be second only to those from the treatment of low-back pain. The response from this survey also suggests that the level of primary care, health promotion and prevention activities of chiropractors surpasses that of other physicians.

Most chiropractors are understandably enthusiastic about maintenance care. After all, it contributes significantly to their income. Statements such as these are typical:

- A healthy spine is one that can function properly and is properly inline to allow the body to function at its optimum capacity. Seeing your chiropractor today will not only allow your body to heal and function better than ever, but also will promote long term health and prevent disease.[3]
- Why do you get your car serviced regularly, even when it's running well? To see that it goes on running well! So once your chiropractor has helped you back to health, don't you owe it to yourself to keep your body in full working order, with regular check-ups?[4]

But do such proclamations stand to reason? Is chiropractic really effective in preventing any disease? My 2009 review evaluated the evidence for or against this approach.[5] No compelling evidence was found to indicate that chiropractic maintenance therapy effectively prevents symptoms or diseases. As SMT has repeatedly been associated with considerable harm (not to mention the cost), the risk/benefit balance of chiropractic maintenance care is not demonstrably positive. Therefore, I concluded, there are no good reasons to recommend it.

A systematic review of 2018 authored by chiropractors agreed. Its authors investigated whether there is any evidence that SMT and chiropractic care can be used in primary prevention or early secondary prevention in diseases other than musculoskeletal conditions.[6] A total of 13 articles were included (8 clinical studies and 5 population studies). Only two clinical studies could be used for data synthesis. None showed an effect of SMT or chiropractic care. The authors concluded that there is *no evidence in the literature of an effect of chiropractic treatment in the scope of primary prevention or early secondary prevention for disease in general.* They furthermore made the following additional comments:

> Beliefs that a spinal subluxation can cause a multitude of diseases and that its removal can prevent them is clearly at odds with present-day concepts, as the aetiology of most diseases today is considered to be multi-causal, rarely mono-causal. It therefore seems naïve when chiropractors attempt to control the combined effects of environmental, social, biological including genetic as

---

[3] https://www.rockymountainchirocare.com/blog/index.php/chiropractic-disease-prevention-not-symptom-prevention-2/.

[4] https://tamworthchiropractic.co.uk/servicing-your-spine-chiropractic-advice/.

[5] Ernst E. Chiropractic maintenance treatment, a useful preventative approach?. *Prev Med*. 2009;49(2–3):99–100. https://doi.org/10.1016/j.ypmed.2009.05.004.

[6] Goncalves G, Le Scanff C, Leboeuf-Yde C. Effect of chiropractic treatment on primary or early secondary prevention: a systematic review with a pedagogic approach. *Chiropr Man Therap*. 2018;26:10. Published 2018 Apr 5. https://doi.org/10.1186/s12998-018-0179-x.

well as noxious lifestyle factors through the simple treatment of the spine. In addition, there is presently no obvious emphasis on the spine and the peripheral nervous system as the governing organ in relation to most pathologies of the human body.

The 'subluxation model' can be summarized through several concepts, each with its obvious weakness. According to the first three, (i) disturbances in the spine (frequently called 'subluxations') exist and (ii) these can cause a multitude of diseases. (iii) These subluxations can be detected in a chiropractic examination, even before symptoms arise. However, to date, the subluxation has been elusive, as there is no proof for its existence. Statements that there is a causal link between subluxations and various diseases should therefore not be made. The fourth and fifth concepts deal with the treatment, namely (iv) that chiropractic adjustments can remove subluxations, (v) resulting in improved health status. However, even if there were an improvement of a condition following treatment, this does not mean that the underlying theory is correct. In other words, any improvement may or may not be caused by the treatment, and even if so, it does not automatically validate the underlying theory that subluxations cause disease…

Despite this embarrassing lack of evidence, many chiropractors continue to promote 'maintenance therapy', i.e. regular sessions of SMT even in the absence of symptoms. One example[7] must stand for the multitude of similar texts promoting maintenance on the Internet:

Maintenance Care is the final stage that ensures that the integrity of the spine is being sustained and supported. Because day to day activities put biomechanical strain on our bodies, we must continue to monitor the health and condition of your spine through periodic evaluations. Regular chiropractic adjustments will help insure you are living at your optimum level of health and function.

The benefits of maintenance care are that minor misalignments can be detected before they become symptomatic or well-established. Maintenance chiropractic care allows for improved posture, enhanced function, better athletic performance, reduced injuries and an overall enjoyable pain-free lifestyle. This type of preventative or wellness care can also save time and money by keeping minor problems from becoming more serious.

Such claims are, however, not based on evidence; my own systematic review[8] concluded that *no compelling evidence was found to indicate that chiropractic*

---

[7] http://mynwwellness.com/blog/5bzybq47w6bk3mdk76stpbk972jljd.

[8] Ernst E. Chiropractic maintenance treatment, a useful preventative approach? Perfusion 2009; 22(5): 164–166.

*maintenance therapy effectively prevents symptoms or diseases. As spinal manip-
ulation has repeatedly been associated with considerable harm, the risk benefit
balance of chiropractic maintenance care is not demonstrably positive. Therefore,
there is no good reason to recommend it.*

In other words, there is no good evidence to suggest that chiropractic disease
prevention is effective. The reason why so many chiropractors nevertheless
recommend it seems rather obvious: it provides a steady income stream for
practitioners who manage to convince their customers that regular treatment
sessions are well worth the effort and expense. Disease prevention by chiropractors will become an even more worrisome topic when, in Chap. 15, we
discuss the attitude of many chiropractors towards immunisation.

Box 13.1 Typical messages about prevention recently picked up on Twitter

- Visit … to learn how chiropractic work can help with disease prevention!
- Check out … most recent interview with … on Chiropractic and disease prevention!
- Find out how chiropractic care helps correct subluxations which is associated with disease prevention.
- Spinal adjustments for disease prevention!
- Many chiropractors seek to care for the whole person from general wellness to disease prevention.
- Find out why chiropractic care is disease prevention rather than disease treatment.
- The cheapest healthcare is to not get sick. Chiropractic is your key to wellness care and disease prevention.
- The practice of Chiropractic is a promotion of health, wellness and disease prevention.
- Chiropractic care focuses on health rather than disease, prevention rather than treatment. Being healthy isn't the absence of disease, it's when your body works correctly so you have optimum physical, mental and social well-being.
- Chiropractic Care reduces stress on the immune system and frees up energy to aid toward disease prevention and maintaining homeostasis.
- Subluxations result in an unbalanced nervous system, discomfort, & disease. Chiropractic care is about health, wellness, & disease prevention.

14

Direct Risks of Spinal Manipulation

Chiropractic is widely recognized 0 as one of the safest drug-free, non-invasive 1
therapies available.
—American Chiropractic Association 2019

Most chiropractors steadfastly deny that SMT can cause serious harm
(Box 14.1). Here is just one of virtually thousands of examples:

> … any concerns about [chiropractic] adjustment are unfounded and merely a
> convenient access point used by a handful of chiropractic detractors or those
> ignorant of the truth.[1]

As evidence for such assertions, chiropractors often rely on a large prospective survey from the UK.[2] It included 50,276 cervical spine manipulations
and noted no serious adverse events (AEs). However, the investigation also
recorded one order of magnitude less minor adverse effects of SMT than
previously published prospective surveys (see below). This casts serious doubts
on its validity. Several explanations of this discrepancy exist:

---

[1] https://www.chiropatient.com/chiropractic-safety/.

[2] Thiel HW, Bolton JE, Docherty S, Portlock JC. Safety of chiropractic manipulation of the cervical
spine: a prospective national survey. *Spine (Phila Pa 1976)*. 2007;32(21):2375–2379. https://doi.org/
10.1097/BRS.0b013e3181557bb1.

© Springer Nature Switzerland AG 2020
E. Ernst, *Chiropractic,*
https://doi.org/10.1007/978-3-030-53118-8_14

- The sample was self-selected.
- It consisted of a relatively small group of participating chiropractors (32% of the total sample).
- The participating chiropractors were more experienced than the average chiropractor (67% had been in practice for 5 or more years).
- They may not always have fully adhered to the protocol of the survey.

The participating chiropractors might, for instance, have employed their experience to select low-risk patients rather than including all consecutive cases, as the protocol demanded. This hypothesis would account for the unusually low rate of minor adverse effects and could explain why no serious complications occurred at all. In any case, given that about 700 serious AEs are on record, their total absence in this survey is surprising.[3]

A 2016 analysis aimed at describing the extent of AE-reporting in published studies of SMT.[4] A total of 368 articles were included in the review. Adverse events were mentioned in only 38% of these articles. The authors concluded that *although there has been an increase in reporting adverse events since the introduction of the 2010 CONSORT guidelines, the current level should be seen as inadequate and unacceptable.*[5]

My team evaluated all 60 RCTs of chiropractic SMT published between 2000 and 2011 and found that half of them did not mention AEs at all. Sixteen RCTs reported that no AEs had occurred (which is hardly credible, since reliable data show that about 50% of patients experience AEs after consulting a chiropractor, see below). Our conclusion was that *adverse effects are poorly reported in recent RCTs of chiropractic manipulations.*

To conduct a clinical trial without mentioning AEs is a violation of medical ethics. The effects of such non-reporting are obvious: anyone looking at the published evidence (for instance via systematic reviews) will get a wrong impression of the safety of SMT. Consequently, chiropractors feel free to claim that very few AEs have been reported, and that therefore SMT is demonstrably safe. Unfortunately, this is precisely what is happening; take, for instance, this statement by the American Chiropractic Association (ACA)[6]:

---

[3] Ernst E. Re: Thiel H W, Bolton J E, Docherty S, et al. Safety of chiropractic manipulation of the cervical spine. Spine 2007;32:2375–8. *Spine (Phila Pa 1976).* 2008;33(5):576–577. https://doi.org/10.1097/BRS.0b013e318165988f.

[4] Gorrell LM, Engel RM, Brown B, Lystad RP. The reporting of adverse events following spinal manipulation in randomized clinical trials-a systematic review. *Spine J.* 2016;16(9):1143–1151. https://doi.org/10.1016/j.spinee.2016.05.018.

[5] Ernst E, Posadzki P. Reporting of adverse effects in randomised clinical trials of chiropractic manipulations: a systematic review. *N Z Med J.* 2012;125(1353):87–140. Published 2012 Apr 20.

[6] https://www.acatoday.org/Patients/Why-Choose-Chiropractic/Chiropractic-Frequently-Asked-Questions.

Chiropractic is widely recognized as one of the safest drug-free, non-invasive therapies available for the treatment of neuromusculoskeletal complaints. Although chiropractic has an excellent safety record, no health treatment is completely free of potential adverse effects …Doctors of chiropractic are well trained professionals who provide patients with safe, effective care for a variety of common conditions…

In the following section, we shall discuss whether such views are truly justified. We will first evaluate the evidence regarding minor AEs and subsequently assess the evidence regarding serious complications of SMT.

Minor Problems

Relatively minor AEs after SMT are extremely common. Our own systematic review of 2002 found that they occur in approximately half of all patients receiving SMT.[7] A more recent study of 771 Finnish patients having chiropractic SMT showed an even higher rate[8]; AEs were reported in 81% of women and 66% of men, and a total of 178 AEs were rated as moderate to severe. Two further studies reported that such AEs occur in 61% and 30% of patients.[9,10] Local or radiating pain, headache, and tiredness are the most frequent adverse effects.

Chiropractors often counter that such AEs are normal events which are necessary steps on the path towards healing. However, in view of the lack of evidence that SMT is an effective treatment for most, if not all conditions (Chaps. 10–13), this is hardly a rational argument. Imagine a drug that has no or very little proven benefit, but causes mild to moderate AEs in about half of all patients. Surely, such a drug would long have been banned from

---

[7] Stevinson C, Ernst E. Risks associated with spinal manipulation. *Am J Med*. 2002;112(7):566–571. https://doi.org/10.1016/s0002-9343(02)01068-9.

[8] Tabell V, Tarkka IM, Holm LW, et al. Do adverse events after manual therapy for back and/or neck pain have an impact on the chance to recover? A cohort study. *Chiropr Man Therap* **27,** 27 (2019). https://doi.org/10.1186/s12998-019-0248-9.

[9] Cagnie B, Vinck E, Beernaert A, Cambier D. How common are side effects of spinal manipulation and can these side effects be predicted?. *Man Ther*. 2004;9(3):151–156. https://doi.org/10.1016/j.math.2004.03.001.

[10] Giles LG. Re: Hurwitz EL, Morgenstern H, Vassilaki M, Chiang L-M. Frequency and clinical predictors of adverse reactions to chiropractic care in the UCLA neck pain study. Spine 2005;30: 1477–84. *Spine (Phila Pa 1976)*. 2006;31(2):250–251. https://doi.org/10.1097/01.brs.0000191703.50217.37.

the market. Or imagine a drug that has been tested in multiple clinical trials of which most fail to even mention AEs.[11]

Whichever way we turn or twist the arguments, one thing is undeniably clear: the popular claim by chiropractors and their supporters that SMT is one of the safest treatments around (like the claim by the ACA cited above) is based on denial, ignorance and wishful thinking. Even more deeply delusional, is the denial that serious complications have been associated with SMT, the issue discussed in the following section.

Cerebral Accidents

If SMT is applied to the neck—which it regularly is, even if the patient is complaining of lower back pain—there is a risk of injury to the fragile structures in this regions. The following case is as good as any of the hundreds of similar reports to illustrate the danger.

In 2015, US doctors published the case of a young woman who developed headache, vomiting, diplopia, dizziness, and ataxia following a neck manipulation by her chiropractor.[12] A CT scan revealed an infarct in the inferior half of the left cerebellar hemisphere and a compression of the fourth ventricle causing moderately severe, acute obstructive hydrocephalus. Magnetic resonance angiography showed severe narrowing and low flow in the intracranial segment of the left distal vertebral artery. The patient was treated surgically and made an excellent recovery. The authors of the case report concluded that *this report illustrates the potential hazards associated with neck trauma, including chiropractic manipulation. The vertebral arteries are at risk for aneurysm formation and/or dissection, which can cause acute stroke.*

Most chiropractors are in denial about this danger and usually quote the Cassidy study[13] which concluded that *vertebral artery accident (VBA) stroke is a very rare event in the population. The increased risks of VBA stroke associated with chiropractic and primary care physician visits is likely due to patients with headache and neck pain from VBA dissection seeking care before their stroke. We found no evidence of excess risk of VBA stroke associated chiropractic care*

---

[11]Ernst E, Posadzki P. Reporting of adverse effects in randomised clinical trials of chiropractic manipulations: a systematic review. *N Z Med J*. 2012;125(1353):87–140. Published 2012 Apr 20.

[12]Jones J, Jones C, Nugent K. Vertebral artery dissection after a chiropractor neck manipulation. *Proc (Bayl Univ Med Cent)*. 2015;28(1):88–90. https://doi.org/10.1080/08998280.2015.11929202.

[13]Cassidy JD, Boyle E, Côté P, et al. Risk of vertebrobasilar stroke and chiropractic care: results of a population-based case-control and case-crossover study [published correction appears in Spine (Phila Pa 1976). 2010 Mar 1;35(5):595]. *Spine (Phila Pa 1976)*. 2008;33(4 Suppl):S176–S183. https://doi.org/10.1097/BRS.0b013e3181644600.

compared to primary care. Yet, the Cassidy study was but one of several case-control studies investigating this subject. And the totality of all such studies does quite clearly suggest that neck manipulation can cause a stroke. More importantly, a re-analysis[14] of the Cassidy study found a classification error in the paper. Therefore, the researchers re-analysed the data, and the re-calculated results indicated a significant risk of SMT. For patients less than 45 years of age, the OR was as high as 6.91. The authors concluded: *If our estimates of case misclassification are applicable outside the VA population, ORs for the association between SMT exposure and CAD are likely to be higher than those reported using the Rothwell/Cassidy strategy, particularly among younger populations.*

Perhaps even more critical is that fact that cases of severe complications after SMT continue to be reported with depressing regularity suggesting that SMT does cause serious damage. Here are but a few recent examples:

- Danish doctors reported a case of a patient with bilateral vertebral artery dissection (VAD) causing embolic occlusion of the basilar artery (BA).[15] The symptoms started after chiropractic SMT of the neck. The patient presented with acute onset of neurological symptoms immediately following SMT in a chiropractic facility. Acute magnetic resonance imaging (MRI) showed ischemic lesions in the right cerebellar hemisphere and occlusion of the cranial part of the BA. Angiography demonstrated bilateral VADs. Symptoms remitted after endovascular therapy, which included dilatation of the left vertebral artery (VA) and extraction of thrombus from the BA. After 6 months, the patient still had minor sensory and cognitive deficits. The authors concluded that *this case underlines the need to suspect VAD in patients presenting with neurological symptoms following SMT*.
- Korean neurosurgeons reported the case of a man with signs of cerebellar dysfunction, vertigo and imbalance. Two weeks before, he had been treated by a chiropractor for neck pain. At the time of admission, brain computed tomography, magnetic resonance imaging, and angiography revealed an acute infarction in the left PICA territory and occlusion of the extracranial vertebral artery as a result of a dissection of his vertebral artery (VA).

[14]Cai X, Razmara A, Paulus JK, et al. Case misclassification in studies of spinal manipulation and arterial dissection. *J Stroke Cerebrovasc Dis.* 2014;23(8):2031–2035. https://doi.org/10.1016/j.jstrok ecerebrovasdis.2014.03.007.

[15]Mikkelsen R, Dalby RB, Hjort N, Simonsen CZ, Karabegovic S. Endovascular Treatment of Basilar Artery Thrombosis Secondary to Bilateral Vertebral Artery Dissection with Symptom Onset Following Cervical Spine Manipulation Therapy. *Am J Case Rep.* 2015;16:868-871. Published 2015 Dec 9. https://doi.org/10.12659/ajcr.895273.

Angiography revealed complete occlusion of the left PICA and arterial dissection was shown in the extracranial portion of the VA. The patient was treated with antiplatelet therapy and discharged three weeks later without any sequelae. The authors concluded that *the possibility of VA dissection should be considered at least once in patients presenting with cerebellar dysfunctions with a recent history of chiropractic cervical manipulation.*

- Dutch neurologists described the case of a woman who presented at their outpatient clinic with a five-week history of severe postural headache, tinnitus and nausea.[16] The symptoms had first occurred soon after chiropractic manipulation of the cervical spine. Cranial MRI showed findings characteristic for intracranial hypotension syndrome. Cervical MRI revealed a large posterior dural tear at the level of C1-2. Following unsuccessful conservative therapy, the patient underwent a lumbar epidural blood patch after which she recovered rapidly. The authors concluded *that manipulation of the cervical spine can cause a dural tear and subsequently an intracranial hypotension syndrome. Postural headaches directly after spinal manipulation should therefore be a reason to suspect this complication.*

- A US neurosurgeon published the case report of a woman with a 3-day history of the acute onset of severe left temporal headache. Her current problem started after an activator treatment (Chap. 5) to the upper cervical spine. Based on MRI characteristics, a haemorrhage was determined to be primarily subarachnoid and displacing but not involving any brain tissue. After a 4-day hospitalization for evaluation and observation, the patient was discharged, neurologically improved to outpatient follow-up. The author concluded that he was *unable to find a single documented case in which a brain hemorrhage in any location was reported from activator treatment. As such, this case appears to represent the first well-documented and reported brain hemorrhage plausibly a consequence of activator treatment. In the absence of any relevant information in the chiropractic or medical literature regarding cerebral hemorrhage as a consequence of activator treatment, this case should be instructive to the clinician who is faced with a diagnostic dilemma and should not forget to inquire about activator treatment as a potential cause of this complication.*

- Neurologists from Qatar published a case report of a man who presented with acute-onset neck pain associated with sudden onset right-sided hemiparesis and dysphasia after chiropractic manipulation for chronic neck

---

[16]Tazelaar GH, Tijssen CC. Liquorhypotensiesyndroom na cervicale manipulatie [Intracranial hypotension syndrome following manipulation of the cervical spine]. *Ned Tijdschr Geneeskd*. 2014;158:A7050.

pain.[17] Magnetic resonance imaging revealed bilateral internal carotid artery dissection and left extracranial vertebral artery dissection with bilateral anterior cerebral artery territory infarctions and large cortical-sparing left middle cerebral artery infarction. The authors concluded that *chiropractic cervical manipulation can result in catastrophic vascular lesions.*

For our own case series of serious AEs after chiropractic, we obtained data on neurological complications of SMT from members of the Association of British Neurologists.[18] They were asked to report cases referred to them of neurological complications occurring within 24 h of cervical spine manipulation over a 12-month period. The response rate was 74%. 24 respondents reported a total of 35 cases. These included:

- 7 cases of stroke in brainstem territory (4 with confirmation of VA dissection),
- 2 cases of stroke in carotid territory,
- 1 case of acute subdural haematoma,
- 3 cases of myelopathy,
- 3 cases of cervical radiculopathy.

A 2017 systematic review identified the characteristics of AEs occurring after cervical spinal manipulation or cervical mobilization.[19] A total of 227 cases were found; 66% of them had been treated by chiropractors. Manipulation was reported in 95% of the cases, and neck pain was the most frequent indication for the treatment. Cervical arterial dissection (CAD) was reported in 57%, and 46% had immediate onset symptoms. The authors of this review concluded that *there seems to be under-reporting of cases. Further research should focus on a more uniform and complete registration of AEs using standardized terminology.*

Even SMT of the thoracic spine is not risk-free: a review analysed reports describing patients who had experienced severe AEs after SMT of their

---

[17]Melikyan G, Kamran S, Akhtar N, Deleu D, Miyares FR. Cortex-sparing infarction in triple cervical artery dissection following chiropractic neck manipulation. *Qatar Med J*. 2016;2015(2):16. Published 2016 Jan 14. https://doi.org/10.5339/qmj.2015.16.

[18]Stevinson C, Honan W, Cooke B, Ernst E. Neurological complications of cervical spine manipulation. *J R Soc Med*. 2001;94(3):107–110. https://doi.org/10.1177/014107680109400302.

[19]Kranenburg HA, Schmitt MA, Puentedura EJ, Luijckx GJ, van der Schans CP. Adverse events associated with the use of cervical spine manipulation or mobilization and patient characteristics: A systematic review [published correction appears in Musculoskelet Sci Pract. 2018 May 18]. *Musculoskelet Sci Pract*. 2017;28:32–38. https://doi.org/10.1016/j.msksp.2017.01.008.

thoracic spine.[20] Ten cases were found. The most frequent AE reported was injury to the spinal cord (7 cases), but pneumothorax/haematothorax (2 cases) and CSF leak secondary to dural sleeve injury (1 case) were also reported. The authors concluded that *serious AEs do occur in the thoracic spine, most commonly, trauma to the spinal cord, followed by pneumothorax.*

In this context, a statement from the American Heart Association and American Stroke Association seems relevant.[21] Here is its abstract in full:

Purpose—Cervical artery dissections (CDs) are among the most common causes of stroke in young and middle-aged adults. The aim of this scientific statement is to review the current state of evidence on the diagnosis and management of CDs and their statistical association with cervical manipulative therapy (CMT). In some forms of CMT, a high or low amplitude thrust is applied to the cervical spine by a healthcare professional…

Results—Patients with CD may present with unilateral headaches, posterior cervical pain, or cerebral or retinal ischemia (transient ischemic or strokes) attributable mainly to artery–artery embolism, CD cranial nerve palsies, oculosympathetic palsy, or pulsatile tinnitus. Diagnosis of CD depends on a thorough history, physical examination, and targeted ancillary investigations. Although the role of trivial trauma is debatable, mechanical forces can lead to intimal injuries of the vertebral arteries and internal carotid arteries and result in CD. Disability levels vary among CD patients with many having good outcomes, but serious neurological sequelae can occur. No evidence-based guidelines are currently available to endorse best management strategies for CDs. Antiplatelet and anticoagulant treatments are both used for prevention of local thrombus and secondary embolism. Case-control and other articles have suggested an epidemiologic association between CD, particularly vertebral artery dissection, and CMT. It is unclear whether this is due to lack of recognition of preexisting CD in these patients or due to trauma caused by CMT. Ultrasonography, computed tomographic angiography, and magnetic resonance imaging with magnetic resonance angiography are useful in the diagnosis of CD. Follow-up neuroimaging is preferentially done with noninvasive modalities, but we suggest that no single test should be seen as the gold standard.

Conclusions—CD is an important cause of ischemic stroke in young and middle-aged patients. CD is most prevalent in the upper cervical spine and can involve the internal carotid artery or vertebral artery. Although current biomechanical evidence is insufficient to establish the claim that CMT causes

---

[20] Puentedura EJ, O'Grady WH. Safety of thrust joint manipulation in the thoracic spine: a systematic review. *J Man Manip Ther.* 2015;23(3):154–161. https://doi.org/10.1179/2042618615Y.0000000012.

[21] Biller J, et al. Cervical Arterial Dissections and Association With Cervical Manipulative Therapy. *Stroke*, Vol 45, No 10 (2014). https://doi.org/10.1161/STR.0000000000000016.

CD, clinical reports suggest that mechanical forces play a role in a considerable number of CDs and most population controlled studies have found an association between CMT and VAD stroke in young patients. Although the incidence of CMT-associated CD in patients who have previously received CMT is not well established, and probably low, practitioners should strongly consider the possibility of CD as a presenting symptom, and patients should be informed of the statistical association between CD and CMT prior to undergoing manipulation of the cervical spine.

Collectively the above evidence leaves little doubt that SMT can cause serious complications. Yet, none of the above-cited papers addressed the question as to how frequently they occur. The purpose of this study was to fill this gap.[22] The authors identified cases through a retrospective chart review of patients seen between April 2008 and March 2012 who had a diagnosis of cervical artery dissection following a recent chiropractic manipulation. Relevant imaging studies were reviewed to confirm the findings of a cervical artery dissection and stroke. The authors also conducted telephone interviews with each patient to ascertain the presence of residual symptoms in the affected patients. Of the 141 patients with cervical artery dissection, 12 had documented chiropractic neck manipulation prior to the onset of the symptoms that led to medical presentation. These 12 patients had a total of 16 cervical artery dissections. All 12 patients developed symptoms of acute stroke, confirmed with magnetic resonance imaging or computerized tomography. Follow-up information could be obtained from 9 patients, 8 of whom had residual symptoms and one patient had died. The Tables 14.1 and 14.2 provide further details.

The authors concluded that *in this case series, 12 patients with newly diagnosed cervical artery dissection(s) had recent chiropractic neck manipulation. Patients who are considering chiropractic cervical manipulation should be informed of the potential risk and be advised to seek immediate medical attention should they develop symptoms.*

---

[22]Kennell KA, Daghfal MM, Patel SG, DeSanto JR, Waterman GS, Bertino RE. Cervical artery dissection related to chiropractic manipulation: One institution's experience. *J Fam Pract.* 2017;66(9):556–562.

Table 14.1 Demographics, original symptoms, and frequency of chiropractor use

Case #	Sex/age	Original symptoms[a]	Frequency of chiropractor use	Time from manipulation to development of new symptoms
1	M/32	Chronic neck pain	Occasional	Immediate
2	F/37	Chronic neck pain	Once a month for 15 years	Immediate
3	F/40	Neck pain for a few weeks	Several visits subsequent to onset of pain	Immediate
4	F/22	Motor-vehicle accident (MVA) one month earlier; neck pain and stiffness	Unknown	Immediate
5	F/30	Postpartum neck pain and stiffness	First time	Immediate
6	F/45	Chronic headaches	Regular	Immediate
7	M/45	Chronic neck and back pain post MVA 19 years ago	Occasional	Immediate
8	M/44	Neck stiffness for 2 days; no history of trauma	Unknown	Immediate
9	F/46	Sore neck for a few days	Once or twice a year for past 5 years	Immediate
10	F/27	Chronic neck pain and headaches several years post MVA	First time	Immediate
11	F/29	Neck stiffness and migraines	Once a week for 10 years	2 days
12	F/36	Neck pain and migraines	Regular	2–3 days

[a]Symptoms that led to chiropractic manipulation

Eye Injuries

In 2005, I published a systematic review of ophthalmic AEs after SMT.[23] At the time, there were 14 published case reports. Clinical symptoms and signs included:

---

[23]Ernst E. Ophthalmological adverse effects of (chiropractic) upper spinal manipulation: evidence from recent case reports. *Acta Ophthalmol Scand*. 2005;83(5):581–585. https://doi.org/10.1111/j.1600-0420.2005.00488.x.

Table 14.2 Timing of events and outcomes following chiropractic care

Case #	Post chiropractic manipulation symptoms that led patient to seek medical care	Time from symptom onset to medical visit	Outcome[a]
1	HA, ear and forehead pain, N/V, and blurry vision	Several weeks	Death
2	N/V, dizziness, and visual disturbance	Immediate	Disequilibrium, stubs toes, clumsiness
3	Sensation of pop and onset of neck pain, HA, N/V	Immediate	Unknown
4	New neck pain, N/V	Immediate	Unknown
5	Blurred vision, difficulty speaking and swallowing, right facial paresthesias, and vertigo.	Immediate	Unsteady when eyes closed, right eyelid droop, HAs, dizziness
6	Vertigo and nausea	Immediate	No residual symptoms
7	Visual field defect, nausea, and dizziness	Immediate	Bilateral visual field defects, HAs
8	Weakness in all 4 extremities with numbness, neck pain, and severe posterior HA	Immediate	At 5–7 months post dissection: spasticity of right hand, reduced use of right arm and leg, neck pain, depression, ataxia
9	Neck pain, mild dizziness, and nausea. Two days later, severe eye pain, slurred speech, and syncope	2 days	Left foot weakness, bilateral visual field defects, balance problems requiring use of a cane
10	N/V and severe vertigo	Immediate	Slight limp
11	N/V, near syncope, vertigo, and visual disturbance	Immediate	HAs, left arm weakness
12	Left-sided numbness, clumsiness, tingling, and HA	Immediate	Unknown

HA, headache; N/V, nausea and vomiting
[a]All outcomes >1 year post event unless otherwise indicated

- central retinal artery occlusion,
- nystagmus,
- Wallenberg syndrome,
- ptosis,
- loss of vision,
- ophthalmoplegia,
- diplopia,
- Horner's syndrome.

In most cases, the underlying mechanism was arterial wall dissection. The eventual outcome varied and often included permanent deficits. Causality was frequently deemed likely or certain. I concluded *that upper spinal manipulation is associated with ophthalmological adverse effects of unknown frequency. Ophthalmologists should be aware of its risks. Rigorous investigations must be conducted to establish reliable incidence figures.*

Since the publication of this paper, new evidence has emerged.

- A 2018 case report told the story of a man with left-sided weakness after a syncopal episode.[24] He had been treated with regular chiropractic neck manipulations over the past seven years. His last session had been one month prior to presentation. The patient developed a headache, anisocoria, and ptosis of his right upper eyelid. Computed tomography angiography (CTA) showed an internal carotid occlusion with right middle cerebral artery zone of ischemia, and tissue plasminogen activator (tPA) was administered. Subsequently, the patient experienced vision loss in his right eye. MRI and CTA were repeated, revealing a right ICA dissection from below the ophthalmic artery to the posterior communicating artery. A diagnosis of ophthalmic artery occlusion was made, and he was discharged on anticoagulant therapy. Three months after presentation, vision had improved to light perception, and remains stable at one year after the dissection. The authors of this case report concluded *that internal carotid artery dissection in this case was permanently devastating to the vision of a previously healthy young patient.*
- In 2018, US ophthalmologists published the case of a woman with the acute, painless constant appearance of three spots in her vision immediately after a chiropractor performed cervical spinal manipulation using

---

[24]Regan KA, Youn TS, Iyer SSR. Ophthalmic artery occlusion after chiropractic neck manipulation. *Acta Ophthalmol.* 2018;96(5):e663–e664. https://doi.org/10.1111/aos.13738.

the high-velocity, low-amplitude technique.[25] She had noted the first spot while driving home immediately following a chiropractor neck adjustment, and became aware that there were two additional spots the following day. Slit lamp examination of the right eye demonstrated multiple unilateral pre-retinal haemorrhages with three present inferiorly along with a haemorrhage over the optic nerve and a shallow, incomplete posterior vitreous detachment. The haemorrhages resolved within two months. The authors concluded that *chiropractor neck manipulation has previously been reported leading to complications related to the carotid artery and arterial plaques. This presents the first case of multiple unilateral pre-retinal haemorrhages immediately following chiropractic neck manipulation. This suggests that chiropractor spinal adjustment can not only affect the carotid artery, but also could lead to pre-retinal haemorrhages.*

Other Serious Complications

Vascular accidents are the most frequent serious AEs after chiropractic SMT, but they are certainly not the only complications that have been reported. Other AEs include:

- atlantoaxial dislocation,
- cauda equina syndrome,
- cervical radiculopathy,
- diaphragmatic paralysis,
- disrupted fracture healing,
- dural sleeve injury,
- haematoma,
- haematothorax,
- haemorrhagic cysts,
- muscle abscess,
- muscle abscess,
- myelopathy,
- neurologic compromise,
- oesophageal rupture
- pneumothorax,
- pseudoaneurysm,

[25]Paulus YM, Belill N. Preretinal hemorrhages following chiropractor neck manipulation. *Am J Ophthalmol Case Rep.* 2018;11:181–183. Published 2018 Apr 19. https://doi.org/10.1016/j.ajoc.2018.04.017.

- soft tissue trauma,
- spinal cord injury,
- vertebral disc herniation,
- vertebral fracture.

In the following section, we will look at some of the recently reported problems in more detail.

- Neurosurgeons published the case of a man with a two-week history of right-sided neck pain and tenderness, accompanied by tingling in the hand.[26] The doctor referred the patient to a chiropractor who performed plain X-rays which allegedly showed "mild spasm". The chiropractor then manipulated the patient's neck on two successive days. By the morning of the 3rd visit, the patient reported extreme pain and difficulty walking. Without performing a new examination, the chiropractor manipulated the patient's neck for a third time whereupon the patient immediately became quadriplegic. Despite undergoing an emergency C5 C6 anterior cervical discectomy/fusion to address a massive disc prolapse found on the magnetic resonance scan, the patient remained quadriplegic.
- Canadian chiropractors reported 6 cases that occurred between 2000 and 2022 where Canadian chiropractors were sued for causing or aggravating lumbar disc herniation after SMT.[27] The following conclusions from Canadian courts seem relevant: (1) Informed consent is an on-going process that cannot be entirely delegated to office personnel. (2) When the patient's history reveals risk factors for lumbar disc herniation the chiropractor has the duty to rule out disc pathology as an aetiology for the symptoms presented by the patients before beginning anything but conservative palliative treatment. (3) Lumbar disc herniation may be triggered by spinal manipulative therapy on vertebral segments distant from the involved herniated disc such as the thoracic spine. In this context, it seems

[26] Epstein NE, Forte Esq CL. Medicolegal corner: Quadriplegia following chiropractic manipulation. *Surg Neurol Int*. 2013;4(Suppl 5):S327–S329. Published 2013 May 28. https://doi.org/10.4103/2152-7806.112620.

[27] Boucher P, Robidoux S. Lumbar disc herniation and cauda equina syndrome following spinal manipulative therapy: a review of six court decisions in Canada. *J Forensic Leg Med*. 2014;22:159–169. https://doi.org/10.1016/j.jflm.2013.12.026.

worth mentioning that disc herniations after chiropractic SMT have been reported regularly and for many years.[28,29,30,31,32]

- American neurologists published a case of a man who consulted a chiropractor for his neck pain who treated him with neck SMT.[33] This resulted in a bilateral diaphragmatic paralysis. (Similar cases have previously been reported with some regularity,[34,35,36,37] and damage to other nerves has also been documented to be a possible complication of SMT.[38,39]) The authors concluded that *physicians must be aware of this complication and should be cautious when recommending spinal manipulation for the treatment of neck pain, especially in the presence of preexisting degenerative disease of the cervical spine.*

- Danish neurologists reported the case of a man with drooping of his right upper eyelid and an ipsilateral contracted pupil, combined with pain, weakness, and numbness in his upper right limb.[40] The patient had sought

[28] Murphy DR. Herniated disc with radiculopathy following cervical manipulation: nonsurgical management. *Spine J.* 2006;6(4):459–463. https://doi.org/10.1016/j.spinee.2006.01.019.

[29] Tomé F, Barriga A, Espejo L. Herniación discal múltiple tras manipulación quiropráctica cervical [Multiple disc herniation after chiropractic manipulation]. *Rev Med Univ Navarra.* 2004;48(3):39–41.

[30] Markowitz HD, Dolce DT. Cauda equina syndrome due to sequestrated recurrent disk herniation after chiropractic manipulation. *Orthopedics.* 1997;20(7):652–653.

[31] Slater RN, Spencer JD. Central lumbar disc prolapse following chiropractic manipulation: a call for audit of 'alternative practice'. *J R Soc Med.* 1992;85(10):637–638.

[32] Richard J. Disk rupture with cauda equina syndrome after chiropractic adjustment. *N Y State J Med.* 1967;67(18):2496–2498.

[33] John S, Tavee J. Bilateral diaphragmatic paralysis due to cervical chiropractic manipulation. *Neurologist.* 2015;19(3):65–67. https://doi.org/10.1097/NRL.0000000000000008.

[34] Kaufman MR, Elkwood AI, Rose MI, et al. Reinnervation of the paralyzed diaphragm: application of nerve surgery techniques following unilateral phrenic nerve injury. *Chest.* 2011;140(1):191–197. https://doi.org/10.1378/chest.10-2765.

[35] Davies SJ. "C3, 4, 5 Keeps the Diaphragm Alive." Is phrenic nerve palsy part of the pathophysiological mechanism in strangulation and hanging? Should diaphragm paralysis be excluded in survived cases?: A review of the literature. *Am J Forensic Med Pathol.* 2010;31(1):100–102. https://doi.org/10.1097/PAF.0b013e3181c297e1.

[36] Merino-Ramírez MA, Juan G, Ramón M, Cortijo J, Morcillo EJ. Diaphragmatic paralysis following minor cervical trauma. *Muscle Nerve.* 2007;36(2):267–270. https://doi.org/10.1002/mus.20754.

[37] Schram DJ, Vosik W, Cantral D. Diaphragmatic paralysis following cervical chiropractic manipulation: case report and review. *Chest.* 2001;119(2):638–640. https://doi.org/10.1378/chest.119.2.638.

[38] Gouveia LO, Castanho P, Ferreira JJ, Guedes MM, Falcão F, e Melo TP. Chiropractic manipulation: reasons for concern?. *Clin Neurol Neurosurg.* 2007;109(10):922–925. https://doi.org/10.1016/j.clineuro.2007.08.004.

[39] Schmidley JW, Koch T. The noncerebrovascular complications of chiropractic manipulation. *Neurology.* 1984;34(5):684–685. https://doi.org/10.1212/wnl.34.5.684.

[40] Foss-Skiftesvik J, Hougaard MG, Larsen VA, Hansen K. Clinical Reasoning: Partial Horner syndrome and upper right limb symptoms following chiropractic manipulation. Neurology May 2015, 84 (21) e175–e180; https://doi.org/10.1212/WNL.0000000000001616.

chiropractic treatment for his back and neck pain. Following manipulations of the thoracic and cervical spine, the pain intensity initially lessened. One hour after the SMT, the patient experienced the eye and upper limb symptoms. A neurologic examination revealed moderate right-sided ptosis and miosis, decreased strength of the intrinsic and opponens muscles of the right hand, and reduced cutaneous sensation corresponding to the T1 dermatome, with inability to discriminate pain and light touch. An MRI of the thoracic spine showed a para-median herniation of the T1-T2 intervertebral disc compressing the right T1 spinal nerve root. The patient received no surgery, and follow-up examination 6 months later revealed near-complete recovery, with only mild paraesthesia in the T1 segment of his right arm and a subtle ptosis remaining.

- Atlantoaxial dislocation is a dislocation of the first and second vertebrae which means that the spinal cord is in danger of being compressed which, in turn, would have devastating consequences. A case report described a man with a history of old cerebellar infarction who presented with acute left hemiplegia after a chiropractic SMT of the neck and back several hours before the symptom onset. Mild hypoesthesia but no speech disturbance, facial palsy, or neck or shoulder pain were observed. Brown-Sequard syndrome (damage to one side of the spinal cord causing paralysis and loss of feeling on one side) subsequently developed with a hypo-aesthetic sensory level below the right C5 dermatome. A magnetic resonance angiography revealed an atlantoaxial dislocation causing upper cervical spinal cord compression. The patient received decompressive surgery, and his muscle power gradually improved, with partial dependency when performing daily living activities two months later. Two months later, his neurological deficits were much improved.

A review summarised cases of cervical spine injury and myelopathy following SMT of the neck.[41] Its authors assessed all patients who had developed neurological symptoms due to cervical spinal cord injury following neck SMT in a spinal unit in a tertiary hospital between the years 2008 and 2018. Patients with vertebral artery dissections were excluded. A total of 4 patients were identified, two men and two women, aged between 32 and 66 years. In three patients, neurological deterioration appeared after chiropractic adjustment and in one patient after 'tuina' therapy (a form of massage

[41] Salame K, Grundshtein A, Regev G, Khashan M, Lador R, Lidar Z. Acute Presentation of Cervical Myelopathy Following Manipulation Therapy. *Isr Med Assoc J.* 2019;21(8):542–545.

used in Traditional Chinese Medicine). The patients had experienced symptoms within one day to one week after neck manipulation. The patients had signs of:

- central cord syndrome,
- spastic quadriparesis,
- spastic quadriparesis,
- radiculopathy and myelomalacia.

Three patients were managed with anterior cervical discectomy and fusion, while one patient declined surgical treatment.

Australian researchers published a systematic review aimed at evaluating all reports of serious AEs following lumbo-pelvic SMT.[42] They identified 41 relevant articles reporting a total of 77 cases consisting of cauda equina syndrome (29 cases); lumbar disk herniation (23 cases); fracture (7 cases); haematoma or haemorrhagic cyst (6 cases); and12 cases of neurologic or vascular compromise, soft tissue trauma, muscle abscess formation, disrupted fracture healing, and oesophageal rupture.

Deaths

In 2010, I reviewed all the reports of deaths after chiropractic treatments published in the medical literature.[43] My article covered 26 fatalities but it is important to stress that many more might have remained unpublished. The cause usually was a vascular accident involving the dissection of a vertebral artery (see above). The review also makes the following important points:

- … numerous deaths have been associated with chiropractic. Usually high-velocity, short-lever thrusts of the upper spine with rotation are implicated. They are believed to cause vertebral arterial dissection in predisposed individuals which, in turn, can lead to a chain of events including stroke and death. Many chiropractors claim that, because arterial dissection can also occur spontaneously, causality between the chiropractic intervention and arterial dissection is not proven. However, when carefully evaluating

[42]Hebert JJ, Stomski NJ, French SD, Rubinstein SM. Serious Adverse Events and Spinal Manipulative Therapy of the Low Back Region: A Systematic Review of Cases. *J Manipulative Physiol Ther*. 2015;38(9):677–691. https://doi.org/10.1016/j.jmpt.2013.05.009.
[43]Ernst E. Deaths after chiropractic: a review of published cases. *Int J Clin Pract*. 2010;64(8):1162–1165. https://doi.org/10.1111/j.1742-1241.2010.02352.x.

the known facts, one does arrive at the conclusion that causality is at least likely. Even if it were merely a remote possibility, the precautionary principle in healthcare would mean that neck manipulations should be considered unsafe until proven otherwise. Moreover, there is no good evidence for assuming that neck manipulation is an effective therapy for any medical condition. Thus, the risk-benefit balance for chiropractic neck manipulation fails to be positive.

- Reliable estimates of the frequency of vascular accidents are prevented by the fact that under-reporting is known to be substantial. In a survey of UK neurologists, for instance, under-reporting of serious complications was 100%. Those cases which are published often turn out to be incomplete. Of 40 case reports of serious adverse effects associated with spinal manipulation, nine failed to provide any information about the clinical outcome. Incomplete reporting of outcomes might therefore further increase the true number of fatalities.

- This review is focussed on deaths after chiropractic, yet neck manipulations are, of course, used by other healthcare professionals as well. The reason for this focus is simple: chiropractors are more frequently associated with serious manipulation-related adverse effects than osteopaths, physiotherapists, doctors or other professionals. Of the 40 cases of serious adverse effects mentioned above, 28 can be traced back to a chiropractor and none to an osteopath. A review of complications after spinal manipulations by any type of healthcare professional included three deaths related to osteopaths, nine to medical practitioners, none to a physiotherapist, one to a naturopath and 17 to chiropractors. This article also summarised a total of 265 vascular accidents of which 142 were linked to chiropractors. Another review of complications after neck manipulations published by 1997 included 177 vascular accidents, 32 of which were fatal. The vast majority of these cases were associated with chiropractic and none with physiotherapy. The most obvious explanation for the dominance of chiropractic is that chiropractors routinely employ high-velocity, short-lever thrusts on the upper spine with a rotational element, while the other healthcare professionals use them much more sparingly.

Another review published in 2012 summarised published cases of injuries associated with cervical manipulation in China.[44] A total of 156 cases were found. They included the following problems:

---

[44]Wang HH, Zhan HS, Zhang MC, Chen B, Guo K. *Zhongguo Gu Shang*. 2012;25(9):730–736.

- syncope (45 cases),
- mild spinal cord injury or compression (34 cases),
- nerve root injury (24 cases),
- ineffective treatment/symptom increased (11 cases),
- cervical spine fracture (11 cases),
- dislocation or semi-luxation (6 cases),
- soft tissue injury (3 cases),
- serious accident (22 cases) including paralysis, death and cerebrovascular accident.

Manipulation including rotation was involved in 42% of all cases. In total, 5 patients died.

Not all fatalities after SMT get reported in medical journals. In fact, most seem to end up in court and are not retrievable via the medical literature. Here are just two recent examples:

- John Lawler died in 2018 after being treated by a chiropractor. The cause of death was a tear and dislocation of the C4/C5 intervertebral disc. The pathologist's report showed that the deceased's ligaments around vertebrae of the upper spine had been ossified, a common abnormality in elderly patients which limits the range of movement of the neck and therefore can be the reason for patients consulting chiropractors. Mr Lawler's chiropractor had failed to obtain adequately informed consent from her patient and Mr Lawler seemed to have been under the impression that the chiropractor, who used the 'Dr' title, was a medical doctor. There is no evidence to assume that the treatment of Mr Lawler's neck would be effective for his pain located in his leg, the reason for the consultation. The chiropractor used an 'activator' a 'drop table' which applies a larger and not well-controlled force (Chap. 5).[45]
- The American model Katie May died in 2016 as the result of visiting a chiropractor for an adjustment, which ultimately left her with a fatal tear to an artery in her neck. According to Wikipedia,[46] Katie tweeted on January 29, 2016, that she had "pinched a nerve in [her] neck on a photoshoot" and "got adjusted" at a chiropractor. She tweeted on January 31, 2016 that she was "going back to the chiropractor tomorrow." On the evening of February 1, 2016, May "had begun feeling numbness in a hand and dizzy" and "called her parents to tell them she thought she was going to pass out."

---

[45] https://edzardernst.com/2019/11/death-by-chiropractic-neck-manipulation-more-details-on-the-lawler-case/.

[46] https://en.wikipedia.org/wiki/Katie_May.

At her family's urging, May went to Cedars Sinai Hospital; she was found to be suffering a "massive stroke." According to her father, she "was not conscious when we got to finally see her the next day. We never got to talk to her again." Life support was withdrawn on February 4, 2016. Katie's death certificate states that she died when a blunt force injury tore her left vertebral artery and cut off blood flow to her brain. It also says the injury was sustained during a "neck manipulation by chiropractor." Her death is listed as accidental.

Relative Safety

DD Palmer was strictly against drugs; in that tradition, today's chiropractors often insist that drugs are the third leading cause of death, a notion that has been debunked many times.[47] They also like to claim that SMT is much safer that other options to treat pain, such as oral OTC pain killers. In 1995, Dabbs and Lauretti reviewed the risks of cervical SMT and compared them to those of non-steroidal, anti-inflammatory drugs (NSAIDs).[48] They concluded that the best evidence indicates that cervical manipulation for neck pain is safer than the use of NSAIDs, by as much as a factor of several hundred times. This article has since become one of the most-quoted paper by chiropractors, and its conclusion is somewhat of a chiropractic mantra which is being repeated ad nauseam. Even the American Chiropractic Association states that *the risks associated with some of the most common treatments for musculoskeletal pain—over-the-counter or prescription nonsteroidal anti-inflammatory drugs (NSAIDS) and prescription painkillers—are significantly greater than those of chiropractic manipulation.*[49]

But how reliable are the conclusions of Dabbs and Lauretti? The most obvious criticism of their paper has already been mentioned: it is outdated; today we know much more about the risks and benefits of both approaches. Equally important is the fact that we still have no surveillance system to monitor the AEs of SMT. Consequently, our data on this issue are woefully incomplete and rely mostly on haphazardly published case reports. Most adverse events remain unpublished and under-reporting is therefore huge.

---

[47] https://www.medscape.com/viewarticle/917696.

[48] Dabbs V, Lauretti WJ. A risk assessment of cervical manipulation vs. NSAIDs for the treatment of neck pain. *J Manipulative Physiol Ther*. 1995;18(8):530–536.

[49] https://www.acatoday.org/Patients/Why-Choose-Chiropractic/Chiropractic-Frequently-Asked-Questions?utm_source=googleplus&utm_medium=social&utm_campaign=Social_SEO.

We have shown that, in our UK survey, it amounted to exactly 100%.[50] To make matters worse, case reports were excluded from the analysis of Dabbs and Lauretti. In fact, they included only articles providing numerical estimates of risk (even reports that reported no AEs at all), the opinion of experts, and a 1993 statistic from a malpractice insurer. None of these sources would lead to reliable incidence figures; they are thus no adequate basis for a comparative analysis. In contrast, NSAIDs have long been subject to proper post-marketing surveillance systems generating realistic incidence figures of AEs which Dabbs and Lauretti were able to use. It is, however, important to note that the figures they did employ were not from patients using NSAIDs for neck pain. Instead they were from patients using NSAIDs for arthritis. Equally important is the fact that they refer to long-term use of NSAIDs, while cervical manipulation is rarely applied long-term. Therefore, the comparison of risks of these two approaches does not seem to be valid.

Moreover, when comparing the risks between cervical manipulation and NSAIDs, Dabbs and Lauretti seemed to have used incidence per manipulation, while for NSAIDs the incidence figures were based on events per patient using these drugs. Similarly, it remains unclear whether the NSAID risk refers only to patients who had used the prescribed dose, or whether over-dosing (a phenomenon that surely is not uncommon with patients suffering from chronic arthritis pain) was included in the incidence figures.

To obtain a fair picture of the risks in a real life situation, one should perhaps also mention that chiropractors often fail to warn patients of the possibility of AEs.[51] With NSAIDs, by contrast, patients have, at the very minimum, the drug information leaflets that do warn them of potential harm in full detail. Finally, one could argue that the effectiveness and costs of the two therapies need careful consideration. The costs for most NSAIDs per day are certainly much lower than those of manipulations. As to the effectiveness of the treatments, it is clear that NSAIDs do effectively alleviate pain,[52] while the evidence is not conclusively positive in the case of cervical SMT.[53]

---

[50]Stevinson C, Honan W, Cooke B, Ernst E. Neurological complications of cervical spine manipulation. *J R Soc Med*. 2001;94(3):107–110. https://doi.org/10.1177/014107680109400302.

[51]Langworthy JM, Cambron J. Consent: its practices and implications in United kingdom and United States chiropractic practice. *J Manipulative Physiol Ther*. 2007;30(6):419–431. https://doi.org/10.1016/j.jmpt.2007.05.002.

[52]Machado GC, Maher CG, Ferreira PH, Day RO, Pinheiro MB, Ferreira ML. Non-steroidal anti-inflammatory drugs for spinal pain: a systematic review and meta-analysis. *Ann Rheum Dis*. 2017;76(7):1269–1278. https://doi.org/10.1136/annrheumdis-2016-210597.

[53]Gross AR, Hoving JL, Haines TA, et al. A Cochrane review of manipulation and mobilization for mechanical neck disorders. *Spine (Phila Pa 1976)*. 2004;29(14):1541–1548. https://doi.org/10.1097/01.brs.0000131218.35875.ed.

In other words, the much-cited paper by Dabbs and Lauretti is out-dated, poor quality, fatally flawed, and heavily biased. It provides no sound basis for an evidence-based judgement on the relative risks of cervical manipulation and NSAIDs. The notion that cervical manipulations are safer than NSAIDs is therefore not based on reliable data. In fact, no sound evidence exists to date in support of such a hypothesis.

To sum up: in this chapter, we have seen that chiropractic SMT can cause a wide range of very serious complications which occasionally can even be fatal. As there is no AE reporting system of such events, nobody can be sure how frequently they occur. Sadly, this does not conclude our discussions about the risks of chiropractic. In the next chapter, we need to evaluate the arguably even more important topic of indirect risks.

Box 14.1 Typical statements about safety recently picked up on Twitter

- Chiropractic care is widely recognized as one of the safest drug-free, non-invasive therapies available for the treatment of most back and neck problems. Spinal adjustments are extremely safe when performed by a licensed chiropractor.
- Chiropractic adjustments are safe and effective for ensuring proper spinal health in small children.
- Chiropractic treatments, called spinal adjustments are comfortable, safe, and highly effective.
- As new moms know, muscle strains are an all-too-real part of pregnancy. Many women find chiropractic care provides tremendous relief, especially from low back pain that occurs. Spinal adjustments are safe for the pregnant woman and her baby.
- If the twig is bent, so grows the tree! Spinal Adjustments are safe, effective, gentle and help us grow healthy.
- The good news is that spinal adjustments are very safe, and often very rewarding for many types of headaches.
- Chiropractic spinal adjustments are proven to be a safe, effective, & affordable treatment option.
- Spinal adjustments or spinal manipulation are very gentle, safe, and effective movements of the spine.

15

Indirect Risks

Compulsory vaccination is an outrage and a gross interference with the liberty of the people in a land of freedom.
—(DD Palmer)

In the previous chapter, we have seen that chiropractic SMT causes mild to moderately severe AEs in about half of all patients as well as, in an unknown percentage of patients, a wide range of serious complications. To this already less than encouraging record, we must unfortunately add something that arguably is even more worrying: the harm many chiropractors cause by giving poor advice to their patients.

Risks of Useless Treatments

Even though these indirect risks of chiropractic are difficult to capture and therefore not well researched, they are very real indeed. In Chap. 7, we discussed the many therapeutic claims chiropractors tend to make without good evidence to support them. Virtually every single patient consulting a 'straight' chiropractor (Chap. 2) will be diagnosed with a subluxation that allegedly needs an 'adjustment'. Whenever patients fall for such claims, they will receive treatments for non-existing conditions, or—much worse—an existing health problem will be treated with a useless intervention. It is thus inevitable that they get harmed:

© Springer Nature Switzerland AG 2020
E. Ernst, *Chiropractic*,
https://doi.org/10.1007/978-3-030-53118-8_15

- financially, because the usually lengthy series of treatments costs money,
- physically, because the patients' condition is not effectively treated and their suffering will therefore continue.

Eventually, most patients will seek effective treatments, one would hope. But, even if they do, valuable time might have been lost for catching a serious disease early. Our undercover investigation demonstrated this danger clearly.[1] We asked 350 UK chiropractors whether they would recommend chiropractic *treatment* for severe asthma. Most chiropractors advised to go ahead with such an approach. Considering that chiropractic is ineffective for asthma,[2] such advice is clearly dangerous, and in some cases, it might even turn out to be lethal.

Risks of Diagnostic Methods

Since 1910, when BJ Palmer introduced X-ray diagnostics to his colleagues, chiropractors are known to overuse this tool which undoubtedly carries considerable risks.[3] Even today, many chiropractors employ imaging techniques to chase after non-existent spinal subluxations (Chap. 4). This review provides a good summary[4]:

> The use of routine spinal X-rays within chiropractic has a contentious history. Elements of the profession advocate for the need for routine spinal X-rays to improve patient management, whereas other chiropractors advocate using spinal X-rays only when endorsed by current imaging guidelines. This review aims to summarise the current evidence for the use of spinal X-ray in chiropractic practice, with consideration of the related risks and benefits. Current evidence supports the use of spinal X-rays only in the diagnosis of trauma and spondyloarthropathy, and in the assessment of progressive spinal structural deformities such as adolescent idiopathic scoliosis. MRI is indicated to diagnose serious pathology such as cancer or infection, and to assess the

---

[1]Schmidt K, Ernst E. Are asthma sufferers at risk when consulting chiropractors over the Internet?. *Respir Med*. 2003;97(1):104–105. https://doi.org/10.1053/rmed.2002.1404.

[2]Balon J, Aker PD, Crowther ER, et al. A comparison of active and simulated chiropractic manipulation as adjunctive treatment for childhood asthma. *N Engl J Med*. 1998;339(15):1013–1020. https://doi.org/10.1056/NEJM199810083391501.

[3]Ernst E. Chiropractors' use of X-rays. *Br J Radiol*. 1998;71(843):249–251. https://doi.org/10.1259/bjr.71.843.9616232.

[4]Jenkins HJ, Downie AS, Moore CS, French SD. Current evidence for spinal X-ray use in the chiropractic profession: a narrative review. *Chiropr Man Therap*. 2018;26:48. Published 2018 Nov 21. https://doi.org/10.1186/s12998-018-0217-8.

need for surgical management in radiculopathy and spinal stenosis. Strong evidence demonstrates risks of imaging such as excessive radiation exposure, overdiagnosis, subsequent low-value investigation and treatment procedures, and increased costs. In most cases the potential benefits from routine imaging, including spinal X-rays, do not outweigh the potential harms. The use of spinal X-rays should not be routinely performed in chiropractic practice, and should be guided by clinical guidelines and clinician judgement.

Despite this scientific consensus, many chiropractors refuse to comply. A 2018 review, for instance, stated that *to remedy spine-related problems, assessments of X-ray images are essential to determine the spine and postural parameters. Chiropractic/manual therapy realignment of the structure of the spine can address a wide range of pain, muscle weakness, and functional impairments.*[5] Another author explained that *the continuing use of the subluxation paradigm for radiography by chiropractors has had a lingering effect on the profession, a metaphorical hangover of vitalism that is not consistent with modern healthcare practice. As a result of this conflict, arguments within the profession on the use of X-rays contribute to the continuing schism between evidence-based and subluxation-based chiropractors.*[6]

But the risk of diagnostic methods is not confined to X-rays. Chiropractic has a colourful history of using bogus diagnostics. In the early 1920s, BJ Palmer started marketing the neurocalometer (NCM).[7] He claimed that the NCM was a very delicate, sensitive instrument and that, when placed upon the spine:

1. It verifies the proper places for adjustments.
2. It measures the specific degree of vertebral pressures upon nerves.
3. It measures the specific degree of interference to transmission of mental impulses as a result of vertebral pressure.
4. It proves the exact intervertebral foramina that contains bone pressure upon nerves.
5. It detects when the pressure has been released upon nerves at a specific place.

[5] Oakley PA, Cuttler JM, Harrison DE. X-Ray Imaging is Essential for Contemporary Chiropractic and Manual Therapy Spinal Rehabilitation: Radiography Increases Benefits and Reduces Risks. *Dose Response*. 2018;16(2):1559325818781437. Published 2018 Jun 19. https://doi.org/10.1177/155932 5818781437.

[6] Young KJ, Bakkum BW, Siordia L. The Hangover: The Early and Lasting Effects of the Controversial Incorporation of X-Ray Technology into Chiropractic. *Health History*. 2016;18(1):111–136. https://doi.org/10.5401/healthhist.18.1.0111.

[7] Keating JC Jr. Introducing the Neurocalometer: a view from the Fountain Head. *J Can Chiropr Assoc*. 1991;35(3):165–178.

6. It indicates how much pressure was released, if any.
7. It verifies the differences between cord pressure or spinal nerve pressure cases.
8. It establishes which cases we can take and which we should leave alone.
9. It proves by an established record which you can see thereby eliminating all guesswork on diagnoses.
10. It establishes, from week to week, whether you are getting well or not.
11. It makes possible a material reduction in time necessary to get well, thus making health cheaper….

BJ was everything but shy about the NCM: *Along comes the Neurocalometer. You hear me tell much good about it. You hear me say that it is "THE MOST VALUABLE INVENTION OF THE AGE BECAUSE IT PICKS, PROVES AND LOCATES THE CAUSE OF ALL DIS-EASES OF THE HUMAN RACE.*

He more or less forced his followers to lease the NCM at exorbitant costs and insisted that his claims were based on extensive scientific research: *Experimental work on approximately a thousand cases had proven there are many subluxations in the spine which the X-Ray does not locate, causing pressure upon nerves. This instrument locates them. Experimental work also shows that by using the instrument as a check, results can be obtained in from one-fourth to one-half the time now necessary under the present method. In other words, should it take 100 adjustments to get a case well now, it would take only 25 to 50 to get the same case well using the new NEUROCALOMETER. So superior was the device that even BJ himself could not find subluxations as accurately as the NCM:… Eighteen months of education when focalized down to a pin point means where to pick majors and why… in 30 min the Neurocalometer can do more in picking correct majors than anybody attending school for 17 months, or more than I can do after 28 years.* The truth, however, was that the NCM measured precisely nothing of value.

All diagnostic tests have to fulfil certain criteria in order to be useful. The tests used by chiropractors for diagnosing spinal problems have rarely been submitted to proper validations. One such test is Kemp's test, a manual test used by most chiropractors to diagnose problems with lumbar facet joints. The chiropractor rotates the torso of the patient, while her pelvis is fixed; if manual counter-rotative resistance on one side of the pelvis by the chiropractor causes lumbar pain, it is interpreted as a sign of lumbar facet joint dysfunction which would in turn then be treated with SMT.

The objective of this study[8] was to evaluate the existing literature regarding the accuracy of Kemp's test in the diagnosis of facet joint pain compared to a reference standard. All diagnostic accuracy studies comparing the Kemp's test with an acceptable reference standard were located and included in the review. Subsequently, all studies were scored for quality and internal validity. Five articles met the inclusion criteria. Only two studies had a low risk of bias, and three had a low concern regarding applicability. Pooling of data from studies using similar methods revealed that the test's negative predictive value was the only diagnostic accuracy measure above 50% (56.8%, 59.9%). The authors concluded that *currently, the literature supporting the use of the Kemp's test is limited and indicates that it has poor diagnostic accuracy. It is debatable whether clinicians should continue to use this test to diagnose facet joint pain.*

The problem with chiropractic diagnostic methods is not confined to Kemp's test, but extends to most tests employed by chiropractors. If diagnostic methods are not reliable, they produce either false-positive or false-negative findings. When a false-negative diagnosis is made, the chiropractor might not treat a condition that needs attention. More common in chiropractic routine are false-positive diagnoses. This means chiropractors frequently treat conditions, e.g. subluxations (Chap. 4) which the patient does not have. This, in turn, is not just a waste of money and time but also, if the ensuing treatment is associated with risks (Chap. 14), an unnecessary exposure of patients to getting harmed.

Vaccinations

Chiropractors often advise against effective conventional treatments. DD Palmer's aversion against any pharmacological treatment has already been mentioned. Today's chiropractors are by no means free from this nonsensical attitude. It is best researched in relation to chiropractors' irrational stance towards immunisation; overt anti-vaccination sentiments abound within the chiropractic profession.[9] Despite the irrefutable evidence that vaccinations generate hugely more good than harm, many chiropractors simply do not believe in vaccination, will not recommend it to their patients, place emphasis on risk rather than benefit when advising their patients, and recommend SMT instead which, they falsely claim, strengthens the immune system.

---

[8]Stuber K, Lerede C, Kristmanson K, Sajko S, Bruno P. The diagnostic accuracy of the Kemp's test: a systematic review. *J Can Chiropr Assoc.* 2014;58(3):258–267.

[9]Lawrence DJ. Anti-Vaccination Attitudes within the Chiropractic Profession: Implications for Public Health Ethics. *Topics in Integrative Health Care* 2012, Vol. 3(4) 2012.

Our survey of 2003 investigated the nature of the advice UK chiropractors (and other healthcare professionals) give on measles, mumps and rubella vaccination (MMR) vaccination.[10] Online referral directories listing e-mail addresses of UK chiropractors and private websites were visited. All chiropractors thus located received a letter from a (fictitious) patient asking for advice about the MMR vaccination. After a follow-up letter explaining the investigative nature and aim of this project and offering the option of withdrawal, 26% of all respondents withdrew their answers. In total, 32% of the responses could thus be assessed. The results showed that only three of them advised in favour of MMR vaccination.

For locating the origins of the anti-vaccination attitude within the chiropractic fraternity, we need to look into the history of chiropractic (Box 15.1). DD Palmer, the magnetic healer who 'invented' chiropractic about 120 years ago (Chap. 2), left no doubt about his profound disgust for immunisation: *It is the very height of absurdity to strive to 'protect' any person from smallpox and other malady by inoculating them with a filthy animal poison... No one will ever pollute the blood of any member of my family unless he cares to walk over my dead body...".* Palmer's son, BJ Palmer, provided a more detailed explanation for chiropractors' rejection of immunisation: *Chiropractors have found in every disease that is supposed to be contagious, a cause in the spine. In the spinal column we will find a subluxation that corresponds to every type of disease... If we had one hundred cases of small-pox, I can prove to you, in one, you will find a subluxation and you will find the same condition in the other ninety-nine. I adjust one and return his function to normal... There is no contagious disease... There is no infection... The idea of poisoning healthy people with vaccine virus... is irrational. People make a great ado if exposed to a contagious disease, but they submit to being inoculated with rotten pus, which if it takes, is warranted to give them a disease.[11]*

A historical analysis published in 2000 provides further details[12]:

Although there is overwhelming evidence to show that vaccination is a highly effective method of controlling infectious diseases, a vocal element of the chiropractic profession maintains a strongly antivaccination bias. Reasons for this are examined. The basis seems to lie in early chiropractic philosophy, which, eschewing both the germ theory of infectious disease and vaccination,

[10]Schmidt K, Ernst E. MMR vaccination advice over the Internet. *Vaccine*. 2003;21(11-12):1044–1047. https://doi.org/10.1016/s0264-410x(02)00628-x.

[11]Ernst E. Chiropractic: a critical evaluation. *J Pain Symptom Manage*. 2008;35(5):544–562. https://doi.org/10.1016/j.jpainsymman.2007.07.004.

[12]Campbell JB, Busse JW, Injeyan HS. Chiropractors and vaccination: A historical perspective. *Pediatrics*. 2000;105(4):E43. https://doi.org/10.1542/peds.105.4.e43.

considered disease the result of spinal nerve dysfunction caused by misplaced (subluxated) vertebrae. Although rejected by medical science, this concept is still accepted by a minority of chiropractors. Although more progressive, evidence-based chiropractors have embraced the concept of vaccination, the rejection of it by conservative chiropractors continues to have a negative influence on both public acceptance of vaccination and acceptance of the chiropractic profession by orthodox medicine.

Today, anti-vaccination sentiments do undoubtedly persist within the chiropractic profession but tend to be expressed in a less abrupt, more politically correct language. Two examples might illustrate this point:

The International Chiropractors Association recognizes that the use of vaccines is not without risk. The ICA supports each individual's right to select his or her own health care and to be made aware of the possible adverse effects of vaccines upon a human body. In accordance with such principles and based upon the individual's right to freedom of choice, the ICA is opposed to compulsory programs which infringe upon such rights. The International Chiropractors Association is supportive of a conscience clause or waiver in compulsory vaccination laws, providing an elective course of action for all regarding immunization, thereby allowing patients freedom of choice in matters affecting their bodies and health.[13]

Vaccines. What are we taught? That vaccines came on the scene just in time to save civilization from the ravages of infectious diseases. That vaccines are scientifically formulated to confer immunity to certain diseases; that they are safe and effective. That if we stop vaccinating, epidemics will return…And then one day you'll be shocked to discover that … your "medical" point of view is unscientific, according to many of the world's top researchers and scientists. That many state and national legislatures all over the world are now passing laws to exclude compulsory vaccines.… Our original blood was good enough. What a thing to say about one of the most sublime substances in the universe. Our original professional philosophy was also good enough. What a thing to say about the most evolved healing concept since we crawled out of the ocean. Perhaps we can arrive at a position of profound gratitude if we could finally appreciate the identity, the oneness, the nobility of an uncontaminated unrestricted nervous system and an inviolate bloodstream. In such a place, is not the chiropractic position on vaccines self-evident, crystal clear, and as plain as the sun in the sky?[14]

---

[13]http://www.chiropractic.org/?p=ica/policies#immunization.

[14]https://planetc1.com/the-chiropractic-position-on-vaccines/.

Influenza kills thousands of people every year, and immunisation could prevent many of these deaths. Those at particularly high risk, e.g. young children, individuals aged 65 and older and people with severe diseases in their medical history, are therefore encouraged to get immunised. Nova Scotia health officials had to issue warnings about some anti-flu vaccine literature being distributed by a chiropractor.[15] The leaflets suggested that flu shots increase the risk of a child ending up in hospital and falling ill with Alzheimer's disease. The chair of the Nova Scotia College of Chiropractors even defended this misinformation and claimed the author of the pamphlet had done his homework. *Chiropractic is really pro information. Look at the positive, look at the negative, look at both sides, get your information and make the appropriate decision that's right for you.* However, Nova Scotia's chief public health officer, insisted that the message was wrong and added that the pamphlet was confusing to the public: *It's discouraging, but unfortunately there are a range of what I call alternative-medicine practitioners who espouse a whole bunch of views which aren't evidence based.*

Sadly, the problem is not confined to North America. A 2017 survey amongst European chiropractors showed its extent.[16] A total of 1322 responses from chiropractors across Europe were categorised as:

- orthodox or mixers (79.9%)
- and unorthodox or straight (20.1%).

The proportion of those respondents disagreeing or strongly disagreeing with the statement "In general, vaccinations have had a positive effect on global public health" was 57% and 4% in unorthodox and orthodox categories respectively. And sadly, the claim that chiropractic SMT will increase immunity thus rendering vaccinations superfluous is not confined to influenza but was also a prominent feature of chiropractic advertising during the Covid-19 pandemic of 2020.[17]

Andrew Wakefield is the UK gastroenterologist who, in 1998, published evidence suggesting that the MMR vaccination was a cause of autism. His research was later discovered to be fraudulent. In 2010, a statutory tribunal of the UK General Medical Council found three dozen charges proved against

---

[15] https://www.cbc.ca/news/canada/nova-scotia/clayton-park-chiropractor-spreading-anti-flu-vaccine-info-1.2806829.

[16] Gíslason HF, Salminen, JK, Sandhaugen L, et al. The shape of chiropractic in Europe: a cross sectional survey of chiropractor's beliefs and practice. *Chiropr Man Therap* **27**, 16 (2019). https://doi.org/10.1186/s12998-019-0237-z.

[17] https://sciencebasedmedicine.org/chiropractors-falsely-claim-they-can-protect-patients-from-corona virus/.

Wakefield, including 4 counts of dishonesty and 12 counts involving the abuse of developmentally delayed children.[18] Consequently, he was struck off the UK medical register and now lives in the US where he, amongst other things, enjoys lecturing to chiropractors on the dangers of vaccination.

When Donald Trump, who seems to share Wakefield's anti-vaccination stance,[19] became president of the US, Wakefield managed to creep back into the limelight. At one of President Trump's inaugural balls, he was quoted contemplating the overthrow of the (pro-vaccine) US medical establishment: *What we need now is a huge shakeup at the Centers for Disease Control and Prevention (CDC)—a huge shakeup. We need that to change dramatically.*[20]

The National Vaccine Information Center (NVIC) is an organisation which seems to support anti-vaxers of various kinds. Officially they try to give the image of being neutral about vaccinations and state that they are *dedicated to the prevention of vaccine injuries and deaths through public education and to defending the informed consent ethic in medicine. As an independent clearinghouse for information on diseases and vaccines, NVIC does not advocate for or against the use of vaccines. We support the availability of all preventive health care options, including vaccines, and the right of consumers to make educated, voluntary health care choices.* The NVIC recently made the following announcement[21]:

> The International Chiropractic Pediatric Association (ICPA), which was founded by Dr. Larry Webster and represents doctors of chiropractic caring for children, has supported NVIC's mission to prevent vaccine injuries and deaths through public education and to protect informed consent rights for more than two decades. ICPA's 2013 issue of Pathways to Family Wellness magazine features an article written by Barbara Loe Fisher on "The Moral Right to Religious and Conscientious Belief Exemptions to Vaccination."
>
> Pathways to Family Wellness is a full-color, quarterly publication that offers parents timely, relevant information about health and wellness options that will help them make conscious health choices for their families. ICPA offers NVIC donor supporters and NVIC Newsletter subscribers a complimentary digital version or print version of Pathways to Family Wellness magazine at a significant discount. Visit the Pathways subscription page and, when checking out in the shopping cart, add the exclusive code: NVIC.

---

[18] https://en.wikipedia.org/wiki/Andrew_Wakefield.

[19] https://edzardernst.com/2016/11/unbelievable-the-trump-wellness-plan/.

[20] https://www.theguardian.com/society/2018/jul/18/how-disgraced-anti-vaxxer-andrew-wakefield-was-embraced-by-trumps-america.

[21] https://www.nvic.org/NVIC-Vaccine-News/August-2013/icpa-offers-nvic-supporters-special-subscription.aspx.

ICPA also has initiated parenting support groups that meet monthly to discuss health and parenting topics. Meetings are hosted by local doctors of chiropractic and the Pathways website features a directory of local groups. ICPA Executive Director Dr. Jeanne Ohm said "We look forward to many more years of collaborating with NVIC to forward our shared goal of enhancing and protecting the ability of parents to make fully informed health and wellness choices for their children."

In summary, the evidence clearly shows that the advice many chiropractors issue can, if followed, be even more harmful than their spinal manipulations. As such, it can hardly be called professional, an aspect that we will scrutinise more closely in the next chapter.

Box 15.1 Quotes about immunisation by the founding father of chiropractic, DD Palmer

- On May 14, 1796, Jenner first committed the crime of vaccination...
- No person is improved by being poisoned by either smallpox or vaccination.
- [Vaccination] is the biggest piece of quackery and criminal outrage ever foisted upon any civilized people. Medical ignorance by which criminal outrages are murdering our children all over this country...
- Vaccination and inoculation are pathological; Chiropractic is physiological.
- Compulsory vaccination is an outrage and a gross interference with the liberty of the people in a land of freedom.

16

Professionalism and Education

We advocate for public statements and claims of effectiveness for chiropractic care that are honest, legal, decent and truthful.
—(World Federation of Chiropractic)

Professionalism[1] can be defined as the sum of the qualities that characterise a profession or a professional person. The qualities required include specialized knowledge, competency, honesty, integrity and accountability. In this chapter, I will discuss to what extend chiropractors and their organisations act professionally.

Regulators

Healthcare professionals must be regulated, and the foremost aim of any regulator must be to protect the public from harm caused by members of that profession. The regulation of chiropractors differs from country to country, of course; in some countries, chiropractors are regulated by statute; in others they self-regulate. In all instances, institutions exist that keep a check on the professionals.

---

[1] https://www.wfc.org/website/index.php?option=com_content&view=article&id=533:wfc-releases-new-guiding-principles-document&catid=56:news--publications&Itemid=27&lang=en&fbclid=IwAR26GqFjENyJEfGD9_Pseoo-9jcnX8UlrFKXiyFuIOEUWPbqThO5IL_yvew.

© Springer Nature Switzerland AG 2020
E. Ernst, *Chiropractic*,
https://doi.org/10.1007/978-3-030-53118-8_16

In Canada, chiropractors are regulated by statute. In 2017, it was reported that the Manitoba Chiropractic Health Care Commission, a government agency, had been tasked to review the cost effectiveness of chiropractic services. The commission therefore prepared a report in 2004 for the Manitoba province and the Manitoba Chiropractors Association. Subsequently, this report was kept secret.[2] In fact, its findings were kept under wraps until the CBC obtained a leaked copy in 2017.[3] The CBC then disclosed that the report had made 37 recommendations, including:

- Manitoba Health should limit its funding to chiropractic treatment of acute lower back pain.
- Manitoba Health should provide only limited coverage of the treatment of neck pain because the evidence was *ambiguous or at best weakly supportive* and the treatment carried a *not insignificant safety risk.*
- Manitoba Health should not fund chiropractic treatment for anyone under 18, as the literature does not unequivocally justify the efficacy or safety of such treatment.

The report also challenged claims that chiropractic treatments can address a wide variety of medical conditions. It stated that there was not enough evidence to conclude that chiropractic treatments are effective in treating:

- muscle tension,
- migraines,
- HIV,
- carpal tunnel syndrome,
- gastrointestinal problems,
- infertility,
- cancer,
- any paediatric conditions,
- or as a preventive care treatment.

The report furthermore urged Manitoba Health to establish a monitoring system to keep a closer eye on *the advertising practices of the Manitoba Chiropractors Association and its members to ensure claims regarding treatments are restricted to those for which proof of efficacy and safety exist.* It suggested the government should have regulatory powers over chiropractic advertisements.

---

[2] https://www.cbc.ca/news/canada/manitoba/publicly-funded-chiropractic-care-report-1.4076690.

[3] https://www.cbc.ca/news/canada/manitoba/manitoba-chiropractic-report-withheld-1.4078831.

The report also stated that chiropractors should not own and operate X-ray machines.

This document thus confirms much of what we have discussed in previous chapters. To hide this fact is certainly dishonest, unethical and unprofessional. Yet, it is but one of many examples demonstrating that, instead of protecting the public, regulators often protect the interests of chiropractors.

The General Chiropractic Council (GCC) is the statutory body regulating all chiropractors in the UK. Their foremost aim, they claim, is to ensure the safety of patients undergoing chiropractic treatment. The GCC also alleges to be independent and to protect the health and safety of the public by ensuring high standards of practice in the chiropractic profession. But how much of this is actually happening?

In a 2019 article,[4] the GCC claimed that they are considering a new five-year strategy and formulated key aims. Here is the crucial passage from this document:

A clear strategy is vital but, of course, implementation and getting things changed are where the real work lie. With that in mind, we have a specific business plan for 2019 – the first year of the new strategic plan… This means you'll see some really important changes and benefits including:

- Promote standards: review and improvements to CPD processes, supporting emerging new degree providers, a campaign to promote the public choosing a registered chiropractor
- Develop the profession: supporting and enabling work with the professional bodies
- Investigate and act: a full review of, and changes to, our Fitness to Practice processes to enable a more 'right touch' approach within our current legal framework, sharing more learning from the complaints we receive
- Deliver value: a focus on communication and engagement, further work on our culture, a new website, an upgraded registration database for an improved user experience.

The changes being introduced, backed by the GCC's Council, will have a positive effect. I know Nick, the new Chief Executive and Registrar and the staff team will make this a success. You as chiropractors also have an important role to play – keep engaging with us and take your own action to develop the profession, share your ideas and views as we transform the

---

[4] https://www.gcc-uk.org/news.

organisation, and work with us to ensure we maintain public confidence in the profession of chiropractic.

The document thus describes a strategy aimed at promoting chiropractors regardless of whether they are doing more good than harm. This, however, is neither professional nor is it in line with the GCC's stated aims.

The UK Professional Standards Authority (PSA) has the remit to promote the health, safety and wellbeing of patients, service users and the public by raising standards of regulation and voluntary registration of people working in health and care. In July 2014, the PSA audited all 75 of the disciplinary cases the GCC had closed at the initial stages of its fitness to practise (FTP) during the previous 12 months. Here is a short excerpt from the conclusions of the PSA-report[5]:

> The extent of the deficiencies we found in this audit (as set out in detail above) which related to failures across every aspect of the casework framework, as well as widespread failures to comply with the GCC's own procedures, raises concern about the extent to which the public can have confidence in the GCC's operation of its initial stages FTP process.
>
> In summary, the particular areas of failures/weaknesses identified in our audit include:
>
> Ineffective screening on receipt of 'complaints' and inconsistent completion and updating of risk assessments
>
> Customer service issues, including failing to respond to/acknowledge correspondence promptly, failing to provide clear information about the FTP process and failing to provide updates about progress and outcomes within reasonable timeframes
>
> Inadequate investigation of cases through failures to gather or validate relevant evidence or to do so promptly – sometimes as a result of inconsistent and ineffective use of case plans and case reviews
>
> Deficiencies in the evaluation of information by decision-makers and weaknesses in the reasoning provided for decisions, including failures to address all the relevant allegations and/or reaching decisions on the basis of insufficient evidence
>
> Poor record keeping and various data protection breaches or potential breaches
>
> Ineffective systems for the sharing of relevant information between the Registration and FTP teams, leading to inappropriate action being taken in some cases
>
> Widespread non-compliance with internal guidance and procedures.

---

[5]https://edzardernst.com/2015/02/the-uk-general-chiropractic-council-fit-for-purpose/.

We have also concluded that the steps taken by the GCC, in particular the processes it introduced in its procedure manual in February had not at the time of the audit resulted in consistent improvement in the quality of its casework.

The GCC must protect the health and safety of the public by ensuring high standards of practice in the chiropractic profession. Yet, the conclusions of the PSA audit seem to suggest that the GCC lacks professionalism.

Professional Organisations of Chiropractors

On 8 April 2008, Simon Singh and I published a newspaper article on the evidence for several alternative therapies.[6] Here is what we wrote about chiropractic:

Chiropractors use spinal manipulation to realign the spine to restore mobility. Initial examination often includes X-ray images or MRI scans.

Spinal manipulation can be a fairly aggressive technique, which pushes the spinal joint slightly beyond what it is ordinarily capable of achieving, using a technique called high-velocity, low-amplitude thrust – exerting a relatively strong force in order to move the joint at speed, but the extent of the motion needs to be limited to prevent damage to the joint and its surrounding structures.

Although spinal manipulation is often associated with a cracking sound, this is not a result of the bones crunching or a sign that bones are being put back; the noise is caused by the release and popping of gas bubbles, generated when the fluid in the joint space is put under severe stress.

Some chiropractors claim to treat everything from digestive disorders to ear infections, others will treat only back problems.

DOES IT WORK? There is no evidence to suggest that spinal manipulation is effective for anything but back pain and even then conventional approaches (such as regular exercise and ibuprofen) are just as likely to be effective and are cheaper.

Neck manipulation has been linked to neurological complications such as strokes – in 1998, a 20-year-old Canadian woman died after neck manipulation caused a blood clot which led to stroke. We would strongly recommend physiotherapy exercises and osteopathy ahead of chiropractic therapy because they are at least effective and much safer.

If you do decide to visit a chiropractor despite our concerns and warnings, we very strongly recommend you confirm your chiropractor won't manipulate

---

[6]https://www.dailymail.co.uk/health/article-557946/Are-hoodwinked-alternative-medicine-Two-lea ding-scientists-examine-evidence.html.

your neck. The dangers of chiropractic therapy to children are particularly worrying because a chiropractor would be manipulating an immature spine.

Our article prompted the British Chiropractic Association (BCA), the largest professional organisation for chiropractors in the UK, to send a 'confidential' message to its members. This is what it said:

As we are aware that patients or potential patients of our members will be confronted with questions regarding this article, we have put together some comment and Q&As to assist you.

- Please consider this information as strictly confidential and for your use only.
- Only use this if a patient asks about these specific issues; there is nothing to be gained from releasing any information not asked for.
- Do not duplicate these patient notes and hand out direct to the patient or the media; these are designed for you to use when in direct conversation with a patient.

The BCA will be very carefully considering any questions or approaches we may receive from the press and will respond to them using specially briefed spokespeople. We would strongly advise our members not to speak directly to the press on any of the issues raised as a result of this coverage.

Please note that In the event of you receiving queries from the media, please refer these direct to BCA (0118 950 5950 – Anne Barlow or Sue Wakefield) or Publicasity (0207 632 2400 – Julie Doyle or Sara Bailey).

The following points should assist you in answering questions that patients may ask with regard to the safety and effectiveness of chiropractic care. Potential questions are detailed along with the desired 'BCA response':

o "The Daily Mail article seems to suggest chiropractic treatment is not that effective"

Nothing could be further from the truth. The authors have had to concede that chiropractic treatment works for back pain as there is overwhelming evidence to support this. The authors also contest that pain killers and exercises can do the job just as well. What they fail to mention is that research has shown that this might be the case for some patients, but the amount of time it may take to recover is a lot longer and the chance of re-occurrence of the problem is higher. This means that chiropractic treatment works, gets results more quickly and helps prevent re-occurrence of the problem. Chiropractic is the third largest healthcare profession in the world and in the UK is recognised and regulated by the UK Government.

o "The treatment is described as aggressive, can you explain?"

It is important to say that the authors of the article clearly have no direct experience of chiropractic treatment, nor have they bothered to properly research the training and techniques. Chiropractic treatment can take many forms, depending on the nature of the problem, the particular patient's age and medical history and other factors. The training chiropractors receive is overseen by the government appointed regulator and the content of training is absolutely designed to ensure that an individual chiropractor understands exactly which treatment types are required in each individual patient scenario. Gentle technique, massage and exercise are just some of the techniques available in the chiropractor's 'toolkit'. It is a gross generalisation and a demonstration of lack of knowledge of chiropractic to characterise it the way it appeared in the article.

o "The article talked about 'claims' of success with other problems"

There is a large and undeniable body of evidence regarding the effectiveness of chiropractic treatment for musculoskeletal problems such as back pain. There is also growing evidence that chiropractic treatment can help many patients with other problems; persistent headaches for example. There is also anecdotal evidence and positive patient experience to show that other kinds of problems have been helped by chiropractic treatment. For many of these kinds of problems, the formal research is just beginning and a chiropractor would never propose their treatment as a substitute for other, ongoing treatments.

o "Am I at risk of having a stroke if I have a chiropractic treatment?"

What is important to understand is that any association between neck manipulation and stroke is extremely rare. Chiropractic is a very safe form of treatment.

Another important point to understand is that the treatments employed by chiropractors are statistically safer than many other conservative treatment options (such as ibuprofen and other pain killers with side effects such as gastric bleeding) for mechanical low back or neck pain conditions.

A research study in the UK, published just last year studied the neck manipulations received by nearly 20,000 chiropractic patients. NO SERIOUS ADVERSE SIDE EFFECTS WERE IDENTIFIED AT ALL. In another piece of research, published in February this year, stroke was found to be a very rare event and the risk associated with a visit to a chiropractor appeared to be no different from the risk of a stroke following a visit to a GP.

Other recent research shows that such an association with stroke may occur once in every 5.85 million adjustments.

To put this in context, a 'significant risk' for any therapeutic intervention (such as pain medication) is defined as 1 in 10,000.

Additional info: Stroke is a natural occurring phenomenon, and evidence dictates that a number of key risk factors increase the likelihood of an individual suffering a stroke. Smoking, high blood pressure, high cholesterol and family medical histories can all contribute; rarely does a stroke occur in isolation from these factors. Also, stroke symptoms can be similar to that of upper neck pains, stiffness or headaches, conditions for which patients may seek chiropractic treatment. BCA chiropractors are trained to recognise and diagnose these symptoms and advise appropriate mainstream medical care.

o "Can you tell if I am at risk from stroke?"

As a BCA chiropractor I am trained to identify risk factors and would not proceed with treatment if there was any doubt as to the patient's suitability. Potential risks may come to light during the taking of a case history, which may include: smoking, high cholesterol, contraceptive pill, Blood clotting problems/blood thinning meds, heart problems, trauma to the head etc. and on physical examination e.g. high blood pressure, severe osteoarthritis of the neck, history of rheumatoid arthritis

o "Do you ever tell patients if they are at risk?"

Yes, I would always discuss risks with patients and treatment will not proceed without informed consent.

o "Is it safe for my child to be treated by a chiropractor"

It is a shame that the article so generalises the treatment provided by a chiropractor, that it makes such outrageous claims. My training in anatomy, physiology and diagnosis means that I absolutely understand the demands and needs of spines from the newborn baby to the very elderly patient. The techniques and treatments I might use on a 25 year old are not the same as those I would employ on a 5 year old. I see a lot of children as patients at this clinic and am able to offer help with a variety of problems with the back, joints and muscles. I examine every patient very thoroughly, understand their medical history and discuss my findings with them and their parents before undertaking any treatment.

- Chiropractic is a mature profession and numerous studies clearly demonstrate that chiropractic treatment, including manipulative and spinal adjustment, is both safe and effective.
- Thousands of patients are treated by me and my fellow chiropractors every day in the UK. Chiropractic is a healthcare profession that is growing purely because our patients see the results and GPs refer patients to us because they know we get results!

This article is to promote a book and a controversial one at that. Certainly, in the case of the comments about chiropractic, there is much evidence and research that has formed part of guidelines developed by the Royal Society of General Practitioners, NICE and other NHS/Government agencies, has been conveniently ignored. The statements about chiropractic treatment and technique demonstrate that there has clearly been no research into the actual education that chiropractors in the UK receive – in my case a four year full-time degree course that meets stringent educational standards set down by the government appointed regulator.

When, a few days after our article, Simon published his now famous Guardian comment[7] stating that the BCA "happily promote bogus treatments", he was sued for libel by the BCA. The above "strictly confidential" message reveals the BCA's determination to stamp out criticism and lack of professionalism. As it turned out, they were ill-advised. Not only did they lose their libel suit, but they also dragged chiropractic into a deep crisis.[8]

In 2013, the ROYAL COLLEGE OF CHIROPRACTORS (RCC) was created. It has the following objectives[9]:

- to promote the art, science and practice of chiropractic;
- to improve and maintain standards in the practice of chiropractic for the benefit of the public;
- to promote awareness and understanding of chiropractic amongst medical practitioners and other healthcare professionals and the public;
- to educate and train practitioners in the art, science and practice of chiropractic;
- to advance the study of and research in chiropractic.

In 2017, the RCC published a statement regarding the safety of chiropractic[10]:

Experiencing mild or moderate adverse effects after manual therapy, such as soreness or stiffness, is relatively common, affecting up to 50% of patients. However, such 'benign effects' are a normal outcome and are not unique to chiropractic care.

Cases of serious adverse events, including spinal or neurological problems and strokes caused by damage to arteries in the neck, have been associated with spinal manipulation. Such events are rare with estimates ranging from 1 per 2 million manipulations to 13 per 10,000 patients; furthermore, due to the nature of the underlying evidence in relation to such events (case reports, retrospective surveys and case-control studies), it is very difficult to confirm causation (Swait and Finch 2017).

For example, while an association between stroke caused by vertebral artery damage or 'dissection' (VAD) and chiropractor visits has been reported in a few case-control studies, the risk of stoke has been found to be similar after seeing a primary care physician (medical doctor). Because patients with VAD

---

[7]https://www.theguardian.com/commentisfree/2008/apr/19/controversiesinscience-health.

[8]https://en.wikipedia.org/wiki/British_Chiropractic_Association_v_Singh.

[9]https://edzardernst.com/2013/03/what-is-next-a-royal-college-of-window-salesmen/.

[10]Document available at https://rcc-uk.org/wp-content/uploads/2018/04/Chiropractic_the-facts_v3.pdf.

commonly present with neck pain, it is possible they seek therapy for this symptom from a range of practitioners, including chiropractors, and that the VAD has occurred spontaneously, or from some other cause, beforehand (Biller et al. 2014). This highlights the importance of ensuring careful screening for known neck artery stroke risk factors, or signs or symptoms that there is an ongoing problem, is performed prior to manual treatment of patients (Swait and Finch 2017). Chiropractors are well trained to do this on a routine basis, and to urgently refer patients if necessary.

In the name of accuracy and professionalism, several alterations to this statement seem necessary. Here is my revised version:

> Experiencing mild or moderate adverse effects after chiropractic spinal manipulations, such as pain or stiffness (usually lasting 1–3 days and strong enough to impair patients' quality of life), is very common. In fact, it affects around 50% of all patients.
>
> Cases of serious adverse events, including spinal or neurological problems and strokes often caused by damage to arteries in the neck, have been reported after spinal manipulation. Such events are probably not frequent (several hundred are on record including about 100 fatalities). But, as we have never established proper surveillance systems, nobody can tell how often they occur. Furthermore, due to our reluctance to introduce such surveillance, some of us are able to question causality.
>
> An association between stroke caused by vertebral artery damage or 'dissection' (VAD) and chiropractic spinal manipulation has been reported in about 20 independent investigations. Yet one much-criticised case-control study found the risk of stroke to be similar after seeing a primary care physician (medical doctor). Because patients with VAD commonly have neck pain, it is possible they seek therapy for this symptom from chiropractors, and that the VAD has occurred spontaneously, or from some other cause, beforehand (Biller et al., 2014). Ensuring careful screening for known neck artery stroke risk factors, or signs that there is an ongoing problem would therefore be important (Swait and Finch, 2017). Sadly, no reliable screening tests exist, and neck pain (the symptom that might be indicative of VAD) continues to be one of the conditions most frequently treated by chiropractors.

Individual Chiropractors

In his book, P.H. Long provides evidence for the shockingly high rates of sexual misconduct amongst US chiropractors. He also shows that 'billions of dollars are stolen each year' by chiropractors resulting in higher insurance premiums and taxes.[11] Making false claims is undeniably unprofessional. Yet, undeniably, many chiropractors do just that (Chap. 7).

In 2010, we published an investigation which showed that the majority of chiropractors and their associations in the English-speaking world seem to make therapeutic claims that are not supported by sound evidence.[12] A 2019 survey reached similar conclusions; it determined the frequency, type and nature of at-risk advertising by Australian chiropractors and physiotherapists.[13] Two auditors examined the advertisements by 359 chiropractors for material potentially in breach of the regulatory authorities' advertising guidelines. Two-hundred and fifty-eight (72%) audited chiropractors had breaches of the Advertising Guidelines on their websites and Facebook pages.

According to the US National Practitioner Data Bank, between September 1990 and January 2012, a total of 5,796 chiropractic medical malpractice reports were filed.[14] Common reasons for chiropractic malpractice lawsuits were:

- Chiropractor causes stroke: Numerous cases have been documented in which a patient suffers a stroke after getting his or her neck manipulated or adjusted. Especially forceful rotation of the neck from side to side can overextend an artery that runs along the spine, which can result in a blockage of blood flow to the brain. Strokes are among the most serious medical conditions caused by chiropractic treatment, and can result in temporary or permanent paralysis, and even death.
- Herniated disc following adjustment: Although many patients seek the medical attention of a chiropractor after they have experienced a herniated disc, chiropractors can actually be the cause of the problem. Usually a herniated disc is caused by wear and tear, but a sudden heavy strain, increased pressure to the lower back or twisting motions can cause a sudden herniated disc. The stress that chiropractors exercise in their adjustments have been known to be the root cause of some herniated discs.
- Sexual misconduct: The American Chiropractic Association has assembled a code of ethics "based upon the acknowledgement that the social contract dictates the profession's responsibilities to the patient, the public and the profession." Sexual misconduct is among the top ten reasons that patients

---

[11]Long PH. Chiropractic Abuse: An Insider's Lament. American Council on Science and Health (2013).

[12]Document available at http://www.chirowatch.com/Reviews/10-000%20Chiropractic%20claims.pdf.

[13]Simpson JK. At-risk advertising by Australian chiropractors and physiotherapists. *Chiropr Man Therap* **27,** 30 (2019). https://doi.org/10.1186/s12998-019-0247-x.

[14]https://www.hg.org/legal-articles/chiropractors-and-medical-malpractice-lawsuits-29867.

file lawsuits against chiropractors. Often, chiropractic practices are unfamiliar to many new patients and can be misinterpreted as inappropriate even though they are absolutely normal, so it is important that patients familiarize themselves with common chiropractic methods of healing.

Despite the evidence of such unprofessional behaviour, chiropractors often claim that complaints against them are rare. A retrospective cohort study demonstrates this claim to be wrong; it analysed all formal complaints about all registered chiropractors, osteopaths, and physiotherapists in Australia lodged with health regulators between 2011 and 2016.[15] Concerns about professional conduct accounted for more than half of the complaints. The rate of complaints was higher for chiropractors than osteopaths and physiotherapists (29 vs. 10 vs. 5 complaints per 1000 practice years, respectively). Overall, nearly half of the complaints (48%) involved chiropractors, even though chiropractors make up only 14% of the workforce across these three professions. The authors concluded that *their study demonstrates differences in the frequency of complaints by source, issue and outcome across the chiropractic, osteopathic and physiotherapy professions. Independent of profession, male sex and older age were significant risk factors for complaint in these professions. Chiropractors were at higher risk of being the subject of a complaint to their practitioner board compared with osteopaths and physiotherapists.*

A recent investigation assessed claims reported to the Danish Patient Compensation Association and the Norwegian System of Compensation to Patients related to chiropractic from 2004 to 2012.[16] All compensation claims involving chiropractors reported to one of the two associations between 2004 and 2012 were assessed. The results show that 338 claims were registered between 2004 and 2012 of which 300 were included in the analysis. The authors concluded *that chiropractors in Denmark and Norway received approximately one compensation claim per 100.000 consultations. The approval rate was low across the majority of complaint categories and lower than the approval rates for general practioners and physiotherapists. Many claims can probably be prevented if chiropractors would prioritize informing patients about the normal course of their complaint and normal benign reactions to treatment.* The authors make the following additional comments:

[15]Ryan AT, Too LS, Bismark MM. Complaints about chiropractors, osteopaths, and physiotherapists: a retrospective cohort study of health, performance, and conduct concerns. *Chiropr Man Therap.* 2018;26:12. Published 2018 Apr 12. https://doi.org/10.1186/s12998-018-0180-4.

[16]Jevne J, Hartvigsen J, Christensen HW. Compensation claims for chiropractic in Denmark and Norway 2004-2012. *Chiropr Man Therap.* 2014;22(1):37. Published 2014 Nov 7. https://doi.org/10.1186/s12998-014-0037-4.

A particular concern after cervical SMT is dissection of the vertebral and carotid arteries. Seventeen claims concerning CAD were reported in this data, 14 in Denmark and three in Norway, and 11 of these were approved for financial compensation (64.7% approval rate) representing by far the highest approval rate across all complaint categories… chiropractors generally seem to receive more claims per consultation than GPs and physiotherapists, the approval rate is substantially lower and a similar trend is observed in Norway.

Despite the fact that the UK GCC seems to be more interested in protecting chiropractors than the public (see above), the GCC's 2019 'Fitness to Practice Report'[17] disclosed the following information:

- In 2019 the number of complaints received about chiropractors' fitness to practise increased by 37% from 2018.
- Complaints were made about 79 chiropractors.
- Most complaints are received from patients.
- Most complaints relate to substandard treatment.

From the evidence discussed in this chapter, it seems clear that the professional organisations of chiropractic fail in their main duty to adequately protect the public. They also fail to guide their members through leading by example. The consequence is a widespread lack of professionalism of individual chiropractors. It would thus be fair to say that this has now led to a situation where the lack of professionalism has become endemic in the world of chiropractic.

---

[17] Document available at https://www.gcc-uk.org/assets/publications/FTP_Annual_Report_2019.pdf.

17

Ethical Issues

We believe that patients have a fundamental right to ethical, professional care and the protection of enforceable regulation in upholding good conduct and practice.
—(World Federation of Chiropractic)

Medical[1] ethics comprise a set of rules and principles which are essential for all aspects of healthcare, including, of course, chiropractic. The main issues are:

- Autonomy—patients must have the right to refuse or choose their treatments.
- Beneficence—researchers and clinicians must act in the best interest of the patient.
- Non-maleficence—the expected benefits of interventions must outweigh their risks.
- Justice—the distribution of health resources must be fair.
- Respect for persons—patients must be treated with dignity.
- Truthfulness and honesty—claims must not be misleading.

While all of this has long been standard knowledge in conventional healthcare, it is often neglected in so-called alternative medicine (SCAM). It is

---

[1] https://www.wfc.org/website/index.php?option=com_content&view=article&id=533:wfc-releases-new-guiding-principles-document&catid=56:news--publications&Itemid=27&lang=en&fbclid=IwAR26GqFjENyJEfGD9_Pseoo-9jcnX8UlrFKXiyFuIOEUWPbqThO5IL_yvew.

© Springer Nature Switzerland AG 2020
E. Ernst, *Chiropractic*,
https://doi.org/10.1007/978-3-030-53118-8_17

therefore timely to ask, how much of chiropractic practice abides by the rules of medical ethics?

Previous chapters of this book have focussed on several of the above listed ethical issues (Box 17.1). In this chapter, I will discuss two further areas where chiropractors violate medical ethics with some regularity: nonsensical research projects and informed consent.

Nonsensical Research

At best, nonsensical research would be a waste of precious resources, at worst it violates the beneficence principle; both would be unethical. In chiropractic, nonsensical research—for instance, research that is not based on a plausible hypothesis, or that is a mere cover for promotion—happens far too regularly. If researchers conduct a clinical trial on a research question that is utterly irrational, for example. Here is the abstract of such a trial[2]:

Objective: This study investigated the effect of spinal manipulative therapy (SMT) on the singing voice of male individuals.

Study design: Randomized, controlled, case-crossover trial.

Methods: Twenty-nine subjects were selected among male members of the Heralds of the Gospel. This association was chosen because it is a group of persons with similar singing activities. Participants were randomly assigned to two groups: (A) chiropractic SMT procedure and (B) nontherapeutic transcutaneous electrical nerve stimulation (TENS) procedure. Recordings of the singing voice of each participant were taken immediately before and after the procedures. After a 14-day period, procedures were switched between groups: participants who underwent SMT on the first day were subjected to TENS and vice versa. Recordings were subjected to perceptual audio and acoustic evaluations. The same recording segment of each participant was selected. Perceptual audio evaluation was performed by a specialist panel (SP). Recordings of each participant were randomly presented thus making the SP blind to intervention type and recording session (before/after intervention). Recordings compiled in a randomized order were also subjected to acoustic evaluation.

Results: No differences in the quality of the singing on perceptual audio evaluation were observed between TENS and SMT.

---

[2] https://pubmed.ncbi.nlm.nih.gov/26165173-effect-of-spinal-manipulative-therapy-on-the-singing-voice/?from_term=chiropractic%2C+voice&from_size=10&from_pos=3.

Conclusions: No differences in the quality of the singing voice of asymptomatic male singers were observed on perceptual audio evaluation or acoustic evaluation after a single spinal manipulative intervention of the thoracic and cervical spine.

Nonsensical research usually happens when naïve enthusiasts dabble in science in order to promote their trade without realising that research would require a minimum of expertise, education and training. Scientists know that any reasonable research project requires a reasonable hypothesis. 'Does SMT affect the singing voice of male individuals?', and many other research questions of chiropractic enthusiasts do not belong to this category.

Another way of doing nonsensical research is to run a clinical trial of which the result is obvious before the project even started. 'Pragmatic' trials are currently popular with chiropractors. These studies can be designed in such a way that they will inevitably produce the findings that researchers intended to produce. The 'A+B versus B' (chiropractic plus standard care versus standard care alone) study design is an obvious example of this type of abuse.

This trial, published in the prestigious JAMA, was aimed at determining whether the addition of chiropractic care to usual medical care results in better pain relief and pain-related function when compared with usual medical care alone.[3] It recruited active-duty US service members with low back pain. Usual medical care included self-care, medications, physical therapy, and pain clinic referral. Chiropractic care included spinal manipulative therapy in the low back and adjacent regions and additional therapeutic procedures such as rehabilitative exercise, cryotherapy, superficial heat, and other manual therapies. Results were in favour of usual medical care plus chiropractic care compared with usual medical care alone (see figure).

---

[3]Goertz CM, Long CR, Vining RD, Pohlman KA, Walter J, Coulter I. Effect of Usual Medical Care Plus Chiropractic Care vs Usual Medical Care Alone on Pain and Disability Among US Service Members With Low Back Pain: A Comparative Effectiveness Clinical Trial. *JAMA Netw Open.* 2018;1(1):e180105. https://doi.org/10.1001/jamanetworkopen.2018.0105.

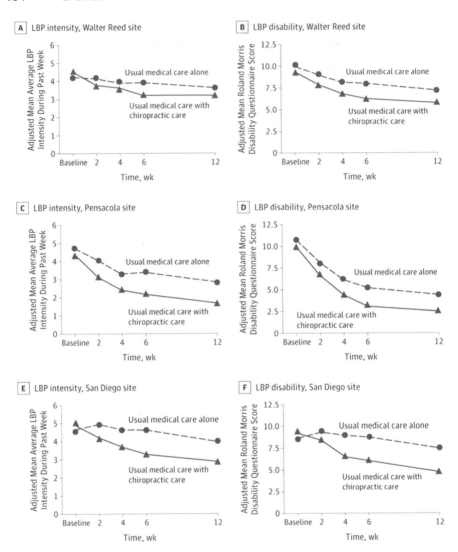

The authors concluded that *chiropractic care, when added to usual medical care, resulted in moderate short-term improvements in low back pain intensity and disability in active-duty military personnel. This trial provides additional support for the inclusion of chiropractic care as a component of multidisciplinary health care for low back pain, as currently recommended in existing guidelines.*

Usual care plus chiropractic SMT is inevitably going to be better that usual care alone, if only for the placebo effect the latter would generate. This becomes evident, if we think of the two interventions as amounts of money: chiropractic care = $10; medical care = $20; 10 + 20 is obviously more than 20.

Study following the 'A+B versus B' design will therefore always generate a positive result. If that is so, it is a waste of time and resources to plan, conduct and publish them. Wasting resources is unethical, and so is misleading everyone into believing that such a study tested the effectiveness of an intervention, while it was merely a smokescreen for producing a pre-determined outcome. The 'A+B versus B' study design cannot control for placebo effects. The results of the study are therefore in perfect agreement with the assumption that chiropractic care is a placebo.

Nonsensical research causes tangible harm. Apart from wasting resources and misleading people, it abuses the willingness of patients to participate in research by telling them that the effort is a worthwhile sacrifice. In reality, it is usually an unethical attempt to generate findings that fool us all. It gives science a bad name and can lead to patients' unwillingness to take part in research that truly does need doing. The damage caused by nonsensical research projects is therefore considerable.

Informed Consent

Informed consent is an essential precondition for any healthcare practice. It requires the clinician or researcher giving the patient full information about the condition and the possible treatments. Amongst other things, the following information may be needed in a standard clinical setting:

- the nature and prognosis of the condition,
- the evidence regarding the efficacy and risks of the proposed treatment,
- the evidence regarding other treatment options.

Depending on the precise circumstances of the clinical situation, consent can be given either in writing or verbally. Not obtaining any form of informed consent is a violation of the most fundamental ethics of health care. In chiropractic, informed consent is often woefully neglected. This may have more than one reason:

- practitioners have frequently no adequate training in medical ethics,
- there is no adequate regulation and control of practitioners,
- practitioners have conflicts of interest and might view informed consent as commercially counterproductive.

Let me explain this using three different scenarios with an asthma patient seeking chiropractic care.

SCENARIO 1
Our patient has experienced breathing problems and has heard that chiropractors might be able to treat this kind of condition. He consults a 'straight' chiropractor, one who adheres to Palmer's gospel of 'subluxation'. She explains to the patient that chiropractors use a holistic approach. By adjusting subluxations in the spine, she is confident to stimulate healing which will naturally ease the patient's breathing problems. No conventional diagnosis is discussed, nor is there any mention of the prognosis, likelihood of benefit, risks of treatment and alternative therapeutic options.

SCENARIO 2
Our patient consults a chiropractor who does not fully believe in the 'subluxation' theory of chiropractic. She conducts a thorough examination of our patient's spine and diagnoses several spinal segments that are blocked. She tells our patient that he might be suffering from asthma and that spinal manipulation might remove the blockages and thus increase the mobility of the spine which, in turn, would alleviate his breathing problems. She does not mention the risks of the proposed interventions nor other therapeutic options.

SCENARIO 3
Our patient visits a chiropractor who considers herself a back pain specialist. She takes a medical history and conducts a physical examination. Subsequently she informs the patient that her breathing problems could be due to asthma and that she is neither qualified nor equipped to ascertain this diagnosis. She tells our patient that chiropractic is not an effective treatment for asthma, but that his GP would be able to firstly make a proper diagnosis and secondly prescribe the optimal treatment for her condition. She writes a short note summarizing her thoughts and hands it to our patient to give it to his GP.

Evidently, the chiropractor in scenario 1 and 2 failed regarding informed consent. Only scenario 3 describes a behaviour that is ethically acceptable. But how likely is scenario 3? The truth is that it is an extremely rare turn of events. Even if well-versed in both medical ethics and scientific evidence, a chiropractor might think twice about providing all the information required for informed consent—because, as scenario 3 demonstrates, informed consent in chiropractic essentially discourages a patient from agreeing to be treated. In other words, chiropractors have a powerful conflict of interest which prevents them from adhering to the rules of informed consent, even if this means ignoring existing guidelines.

In 2016, the General Chiropractic Council (GCC) has issued guidance[4] to its members about informed consent. Here is the relevant passage:

> The information you provide to the patient must be clear, accurate and presented in a way that the patient can understand… Patients must be fully informed about their care. You must not rely on a patient to ask questions about their care, the responsibility to fully inform patients about their care lies with you. When discussing with patients the expected outcomes of their care, chiropractors must fully discuss the risks as well as the benefits and explore with the patient what other factors they may see as relevant to making a decision.
>
> When explaining risks, you must provide the patient with clear, accurate and up-to-date information about the risks of the proposed treatment and the risks of any reasonable alternative options, in a way that the patient can understand. You must discuss risks that occur often, those that are serious even if very unlikely and those that a patient is likely to think are important. You must encourage patients to ask questions, so that you can understand whether they have particular concerns that may influence their decision and you must answer honestly.

This guideline essentially demands that a chiropractor must inform each patient who is about to be treated with SMT—almost all patients consulting chiropractors—that:

- this treatment has not been shown to be effective for non-spinal conditions (Chap. 11),
- for back and neck pain, it might help but not better than other conservative therapies (Chap. 10),
- in about half of all patients, SMT leads to mild to moderate adverse effects that typically last 2–3 days and are severe enough to interfere with the patient's quality of life (Chap. 14),
- in an unknown number of patients, it might lead to severe complications, including stroke and death (Chap. 14),
- there are other options for your problem that are more effective and/or less harmful.

So, have the GCC or any other professional organisations of chiropractors tested how many patients would consent under these conditions? Have they

---

[4]https://www.gmc-uk.org/ethical-guidance/ethical-guidance-for-doctors/consent/part-2-making-decisions-about-investigations-and-treatment.

checked how many chiropractors do actually follow these guidelines? The only study available on this topic is not encouraging; here is the abstract[5]:

> Background: A patient's right to accept or reject proposed treatment is both an ethical and legal tenet. Valid consent is a multifaceted, controversial and often complicated process, yet practitioners are obligated to try to obtain consent from their patients. Its omission is a common basis for malpractice suits and increasing utilization of complementary and alternative services in conventional medical settings is intensifying the focus on medical liability issues. This has important implications for individual professions and their members.
>
> Objective: To investigate approaches to consent among a small (n = 150) sample of practicing UK chiropractors.
>
> Results: Of 150 randomly selected chiropractic practitioners in the United Kingdom, 55% responded. Of these, 25% report not informing patients of physical examination procedures prior to commencement. By contrast, only 6% do not fully explain proposed treatment, although over one-third do not advise patients of alternative available treatments. Nearly two-thirds of the practitioners report that there are no specific procedures for which they always obtain written consent and 18% that there are no instances in which they document when verbal consent has been obtained. Ninety-three percent said they always discuss minor risk with their patients but only 23% report always discussing serious risk. When treatment carries a possible risk of a major side-effect only 14% of the sample obtain formal written consent. Documentation of patient understanding is omitted by 75% of practitioners in this sample.
>
> Conclusion: Results suggest that valid consent procedures are either poorly understood or selectively implemented by UK chiropractors.

The conclusion is as clear as it is worrying: chiropractors are disregarding medical ethics on a daily basis. In fact, it almost seems as though the ethical practice of chiropractic is a contradiction in terms.

Box 17.1 Situations in which chiropractors are likely to violate medical ethics

- Chiropractor diagnoses a spinal subluxation in a patient (subluxations do not exist, telling lies is unethical)
- Chiropractor adjusts a subluxation (unnecessary treatment means an unnecessary expense which is not in the best interest of the patient)

---

[5]Langworthy JM, le Fleming C. Consent or submission? The practice of consent within UK chiropractic. *J Manipulative Physiol Ther.* 2005;28(1):15–24. https://doi.org/10.1016/j.jmpt.2004.12.010.

- Chiropractor fails to obtain informed consent (see above)
- Patient experiences pain due to the adjustment (this might violate the non-maleficence principle)
- The adjustment is not evidence-based and the condition treated by the chiropractor does not improve (this might violate the beneficence principle)
- Chiropractor convinces patient that regular treatments are necessary for staying healthy (this could violate most of the above-listed ethical principles).

18

Postscript

In the introduction, I stated that I wanted you to make up your own mind about the value of chiropractic. To do this, you need reliable information. And to obtain reliable information, you need someone who is well-informed and independent.

What you don't need is information from people who make their living from chiropractic. Remember, it is difficult to get someone to understand something when their salary depends on them not understanding it! Something else you don't need? People who tell you they had a good experience with their chiropractor. Remember, the plural of anecdote is anecdotes, not evidence!

After reading my book, you may feel little enthusiasm for consulting a chiropractor. I am sure that this is not a bad thing—not because I have something against chiropractors personally, but because I am in favour of applying the best available evidence. And the evidence is clear: chiropractic might help a little with back pain, but there are better, less expensive and safer options. For all other conditions, chiropractic is even less useful and possibly even dangerous.

No doubt, my book will infuriate many fans of chiropractic. They will claim that I am incompetent, biased and dishonest. But we have seen that chiropractors make more bogus claims than any other talented snake-oil salesperson could ever dream of. So, I am confident that you will take such personal attacks on me with a pinch of salt.

I have presented you with the evidence, provided the references so that you can double-check and have told you the often-uncomfortable truth. I trust you will use this information wisely and stay healthy.

© Springer Nature Switzerland AG 2020
E. Ernst, *Chiropractic*,
https://doi.org/10.1007/978-3-030-53118-8_18

Glossary

Adjustment Any chiropractic therapeutic procedure that ultimately uses force, leverage, direction, amplitude and velocity, which is applied to specific joints and adjacent tissues. Chiropractors commonly use such procedures to influence joint and neurophysiological function.

Biomechanics The study of structural, functional and mechanical aspects of human motion. It is concerned mainly with external forces of either a static or dynamic nature, dealing with human movement.

Fixation The state whereby an articulation has become fully or partially immobilized in a certain position, restricting physiological movement.

Joint manipulation A manual procedure involving directed thrust to move a joint past the physiological range of motion, without exceeding the anatomical limit.

Joint mobilization A manual procedure without thrust, during which a joint normally remains within its physiological range of motion.

Neuromusculoskeletal Pertaining to the musculoskeletal and nervous systems in relation to disorders that manifest themselves in both the musculoskeletal and nervous systems, including disorders of a biomechanical or functional nature.

Palpation (1) The act of feeling with the hands. (2) The application of variable manual pressure through the surface of the body for the purpose of determining the shape, size, consistency, position, inherent motility and health of the tissues beneath.

© Springer Nature Switzerland AG 2020
E. Ernst, *Chiropractic*,
https://doi.org/10.1007/978-3-030-53118-8

Posture (1) The attitude of the body. (2) The relative arrangement of the parts of the body. Good posture is that state of muscular and skeletal balance that protects the supporting structures of the body against injury or progressive deformity irrespective of the attitude (erect, lying, squatting, stooping) in which the structures are working or resting.

Spinal manipulative therapy Includes all procedures where the hands or mechanical devices are used to mobilize, adjust, manipulate, apply traction, massage, stimulate or otherwise influence the spine and paraspinal tissues with the aim of influencing the patient's health.

Subluxation A lesion or dysfunction in a joint or motion segment in which alignment, movement integrity and/or physiological function are altered, although contact between joint surfaces remains intact. It is essentially a functional entity, which may influence biomechanical and neural integrity.

Subluxation complex (vertebral) A theoretical model and description of the motion segment dysfunction, which incorporates the interaction of pathological changes in nerve, muscle, and ligamentous, vascular and connective tissue.

Thrust The sudden manual application of a controlled directional force upon a suitable part of the patient, the delivery of which effects an adjustment.

Useful Links and Books

The links below will take you to websites that provide further useful information in English about chiropractic.

American Council on Science and Health (ACSH)
Chirobase
Committee for the Scientific Investigation of Claims of the Paranormal (CSICOP)
ebm-first.com
edzardernst.com
Friends of Science in Medicine
https://goodthinkingsociety.org/
James Randi Educational Foundation
The Nightingale Collaboration
PainScience.com
Quackcast
Quackometer
Science-Based Medicine
The SkepDoc
The SkeptVet
What's the Harm?
zenosblog.com.

© Springer Nature Switzerland AG 2020
E. Ernst, *Chiropractic*,
https://doi.org/10.1007/978-3-030-53118-8

Critical Books on Chiropractic

Bonesetting, chiropractic and cultism; A critical study of chiropractic, its history and its methods in its relationship with past and present-day … regarding its status and its future Unknown Binding—1963
by Samuel Homola
https://www.amazon.co.uk/Bonesetting-chiropractic-cultism-relationship-present-day/dp/B0006AYT9W/ref=sr_1_3?dchild=1&keywords=CHIROPRACTIC%2C+CRITICAL&qid=1586765473&s=books&sr=1-3

The Naked Chiropractor: An Insider's Guide to Combating Quackery and Winning the War Against Pain Paperback—28 Oct 2013
by Ph.D. Preston H Long DC
https://www.amazon.co.uk/Naked-Chiropractor-Insiders-Combating-Quackery/dp/1493592823/ref=sr_1_3?dchild=1&keywords=CHIROPRACTIC%2C+PRESTON+LONG&qid=1586765375&s=books&sr=1-3

Lying for Fun and Profit / The Truth about the Media
by Kurt Butler
https://www.smashwords.com/books/view/675547

Inside Chiropractic: A Patient's Guide
by Samuel Homola
https://www.amazon.co.uk/Inside-Chiropractic-Patients-Consumer-Library/dp/1573926981/ref=sr_1_179?dchild=1&keywords=CHIROPRACTIC&qid=1586767772&s=books&sr=1-179

The Chiropractor Hoax: The True Story of Chiropractic Medicine You've Never Been Told (chiropractor abuse, chiropractic abuse, back pain book, treating back pain, trick or treatment, back remedies)
by John Morrison
https://www.amazon.co.uk/Chiropractor-Hoax-Chiropractic-chiropractor-chiropractic-ebook/dp/B07NRHJTQS/ref=sr_1_227?dchild=1&keywords=CHIROPRACTIC&qid=1586768014&s=books&sr=1-227

Chiropractic: The Victim's Perspective
by George Magner
https://www.amazon.co.uk/Chiropractic-Victims-Perspective-Consumer-Library/dp/157392041X/ref=sr_1_1?dchild=1&keywords=Chiropractic%2C+The+Victim%27s+Perspective.&qid=1586762674&s=books&sr=1-1

Chiropractic Abuse: An Insider's Lament Paperback—8 Oct 2013
by Ph.D. Preston H Long D.C

https://www.amazon.co.uk/Chiropractic-Abuse-PhD-Preston-Long/dp/ 0972709495/ref=tmm_pap_swatch_0?_encoding=UTF8&qid=158676 5302&sr=1-1

The Religion of Chiropractic: Populist Healing from the American Heartland
by Holly Folk
https://www.amazon.co.uk/Religion-Chiropractic-Populist-American-Hea rtland/dp/1469632780/ref=sr_1_88?dchild=1&keywords=CHIROPRAC TIC&qid=1586767336&s=books&sr=1-88.

Books by DD Palmer

The chiropractic adjuster: A compilation of the writings of D.D. Palmer
Text-Book of the Science, Art and Philosophy of Chiropractic/The Chiropractor's Adjuster
Chiropractic: A Science, an Art and the Philosophy There of
The Chiropractor: The Philosophy and History of Chiropractic Therapy, Care and Diagnostics by its Founder
The Chiropractor
The science of chiropractic: Its principles and philosophies.